Complex Numbers – The Higher Dimensional Forms

Complex Numbers – The Higher Dimensional Forms

The Natural Algebras
The Nature of Space
The Algebraic Vector Product Spaces
The Abelian Group Algebras
The Unification of Groups, Algebra, and Geometry

Dennis Morris MSc

ABANE & RIGHT

First published in the UK in 2007 by

Abane & Right
31 Long Row
Port Mulgrave
Whitby
TS13 5LF

ISBN 978-0-955600-30-2

Printed in Great Britain

Contents

iv

Historical Note

The association between complex numbers and positions in 2-dimensional euclidean space begins with Wallis (1616 – 1703) in 1673. It was subsequently stated by Wessel (1745 – 1818) in 1798 and again by Gauss (1777 – 1855) in 1799. It was given again by Argand (1768 – 1822) in 1806, and the association subsequently became known as the Argand diagram. In those times, euclidean geometry was very prominent within mathematics and complex numbers were seen as being associated with the euclidean geometry of 2-dimensions. It was inevitable that mathematicians would seek some higher kind of complex numbers associated with 3-dimensional euclidean space, and it was expected that such numbers would exist.

Mathematicians assumed that 2-dimensional space is a sub-space of 3-dimensional space. From this assumption, it follows that the 2-dimensional complex numbers are a sub-algebra of the anticipated 3-dimensional complex numbers and the way forward is to expand the complex numbers upwards into higher dimensions. These efforts were encouraged by some small, but tantalising, success and some almost success. In 1813, Servois[1] proposed a 3-dimensional kind of complex numbers that nearly worked, and in 1843 Hamilton (1805 – 1865) discovered the '4-dimensional' quaternions.

The quaternions work, they include the 2-dimensional complex numbers as a sub-algebra, but they are not an algebraic field. They are a division algebra; they lack multiplicative commutativity. It was immediately obvious to mathematicians of the day that higher dimensional kinds of complex numbers would be multiplicatively non-commutative because rotations within 3-dimensional euclidean space are non-commutative.

Unfortunately, other than the seeming success of the quaternions, all attempts to develop the higher dimensional complex number algebras failed, and so, in the middle of the 19th century, mathematicians developed a new approach to representing space. They turned their attention to describing higher dimensional spaces by use of vectors. Conventional vector algebra is such that 2-dimensional space is a sub-space of 3-dimensional space. Grassman (1809 – 1877) began this approach in 1844, but it was 1862, when he rewrote the work in a more readable form, before his work was appreciated. The vector algebra was expanded and deepened subsequently by Gibbs (1839 – 1903) in 1881 and by Peano (1858 – 1932) in 1888.

Vector algebra underlies large areas of our understanding of the physical universe. It has been very successful and is beloved by physicists and engineers. It is unfortunate that it is not an algebra. From a mathematical point of view, vector algebra is a mess. It is a linear space together with an undeclared metric together with two multiplication operations (the dot product and the cross product), neither of which are a bona-fide algebraic multiplication operation, and, if that's not bad enough, it has no vector product (algebraic multiplication operation).

Grassman was aware of this failing, and, from the start, he sought to introduce the exterior product of two vectors into the vector algebra. In 1878 and 1882 (posthumously) William K. Clifford (1845 – 1879) also attempted to introduce a vector product, known today as the Clifford product, into the vector algebra. In doing so, he invented the Clifford algebras. Unfortunately, neither Grassman's exterior product nor the Clifford

---

[1] Your author is unaware of the years of Servois's birth and death.

product is an algebraic multiplication operation; it is not closed; neither the exterior product of two vectors nor the Clifford product of two vectors is a vector.

Only recently have the bone-fide algebras of the higher dimensional spaces been discovered. These are the natural algebras. They bring with them a radical change in our understanding of the nature of space. They also bring the unification of large areas of mathematics.

Note on the Hurwitz Theorem

The Hurwitz theorem is often stated as:

Hurwitz Theorem:
The only normed algebras over the real field are isomorphic to $\mathbb{R}, \mathbb{C}, \mathbb{H}, \mathbb{K}$.

This is slightly misleading in that this theorem covers only norms that have a euclidean (with all positive signature) quadratic form similar to: $\sqrt{a^2 + b^2}$. In this work, we will find many algebras over the reals that are normed, but the norm will not be a euclidean quadratic form.

Introduction to Section 1

In this section, we discuss a few mathematical concepts. The concepts will often be familiar, but we view them from a different angle. These changes of viewpoint are not changes in any logically deduced mathematics; we are not undoing mathematical proofs. These are changes in viewpoint only. They are changes in the way we see numbers, algebras, and inner products. It is a change of view that is forced upon us by the mathematics that follows in later sections of this work.

The reader might initially feel the changes in viewpoint are unnecessary or unjustified, and we make little attempt to justify them in this section. The necessity of these changes is to be found in the later sections of this work.

Section 1 : Chapter 1

The Matrix Representation of Algebras

The word 'algebra' is, atypically, much abused in modern mathematics to mean almost anything. In this work, it will not be so abused. An algebra is a mathematical construct that satisfies thirteen of the fourteen algebraic field axioms; the exception is multiplicative commutativity. The reader is referred to Appendix 2. Anything less than this is not an algebra. If the mathematical construct satisfies all fourteen of the algebraic field axioms, it is an algebraic field. The algebraic field axioms are the properties of the real numbers, and every algebraic field is a type of number of equal mathematical status to the real numbers.

An algebra is a linear space[2] together with an appropriate multiplication operation. A linear space is a set of linear transformations. Every linear transformation can be represented as a matrix. So every linear space is a particular set of matrices.

We often have the view that a linear space is a set of mathematical objects like real numbers or complex numbers or magic squares. It is. However, a real number or a complex number or a magic square is a linear transformation.

The multiplication operation that we impose upon a linear space to produce an algebra does not have to be matrix multiplication, but, if the imposed multiplication operation is matrix multiplication, then we can do the whole of that algebra in matrix notation. Matrix multiplication seems to be the natural multiplication operation to impose upon a set of matrices. We shall therefore refer to linear spaces that have matrix multiplication for the multiplication operation as the natural algebras.

Nomenclature:

Natural Algebra

An algebra that has matrix multiplication as the multiplication operation.

Natural Algebraic Field

An algebraic field that has matrix multiplication as the multiplication operation.

---

[2] Linear spaces are also known as vector spaces – two names, same thing.

4

    The form of multiplication that we call matrix multiplication is derived from the manipulation of linear equations. Thus, matrix multiplication could be equally well named linear multiplication. Using such parlance, we would define a natural algebra as a linear space together with linear multiplication. We note that linear addition is within the linear space definition.

    Many multiplication operations are matrix multiplication in disguise. For example, multiplication of real numbers is matrix multiplication of 1×1 matrices. The multiplication operation of euclidean complex numbers is also matrix multiplication. We jump the gun to demonstrate[3].

The 2×2 matrix form of a euclidean complex number, $\left(a + \hat{i}b\right)$, is:

$$\begin{bmatrix} a & b \\ -b & a \end{bmatrix}$$

where the elements are all real numbers.
We have:

$$\left(a + \hat{i}b\right)\left(c + \hat{i}d\right) = \left(ac - bd\right) + \left(ad + bc\right)\hat{i}$$

$$\begin{bmatrix} a & b \\ -b & a \end{bmatrix}\begin{bmatrix} c & d \\ -d & c \end{bmatrix} = \begin{bmatrix} ac - bd & ad + bc \\ -\left(ad + bc\right) & ac - bd \end{bmatrix}$$

$$\begin{bmatrix} 0 & 1 \\ -1 & 0 \end{bmatrix}^2 = \begin{bmatrix} -1 & 0 \\ 0 & -1 \end{bmatrix}$$

This is not the only matrix form of the euclidean complex numbers. The 4×4 matrix form with real numbers as elements:

$$\begin{bmatrix} a & 0 & c & 0 \\ 0 & a & 0 & c \\ -c & 0 & a & 0 \\ 0 & -c & 0 & a \end{bmatrix}$$

is also a representation of the euclidean complex numbers. We have:

$$\begin{bmatrix} a & 0 & c & 0 \\ 0 & a & 0 & c \\ -c & 0 & a & 0 \\ 0 & -c & 0 & a \end{bmatrix}\begin{bmatrix} e & 0 & g & 0 \\ 0 & e & 0 & g \\ -g & 0 & e & 0 \\ 0 & -g & 0 & e \end{bmatrix} = \begin{bmatrix} ae - cg & 0 & ag + ce & 0 \\ 0 & ae - cg & 0 & ag + ce \\ -\left(ag + ce\right) & 0 & ae - cg & 0 \\ 0 & -\left(ag + ce\right) & 0 & ae - cg \end{bmatrix}$$

and:

---

[3] Throughout this work, $\hat{i} = \sqrt[2]{-1}$ will have little hat on, as will similar objects.

$$\begin{bmatrix} 0 & 0 & 1 & 0 \\ 0 & 0 & 0 & 1 \\ -1 & 0 & 0 & 0 \\ 0 & -1 & 0 & 0 \end{bmatrix}^2 = \begin{bmatrix} -1 & 0 & 0 & 0 \\ 0 & -1 & 0 & 0 \\ 0 & 0 & -1 & 0 \\ 0 & 0 & 0 & -1 \end{bmatrix}$$

We emphasize that part of the definition of the algebra is that the elements within the matrix are real numbers.

An Aside:

In the case where matrix elements are complex numbers, the block multiplication property of matrices allows us to write the matrix as a larger matrix of real numbers. For example, consider the matrix:

$$\begin{bmatrix} a+\hat{i}b & c+\hat{i}d \\ e+\hat{i}f & g+\hat{i}h \end{bmatrix}$$

We can write this 2×2 matrix as a 4×4 matrix with real numbers as elements:

$$\begin{bmatrix} \begin{bmatrix} a & b \\ -b & a \end{bmatrix} & \begin{bmatrix} c & d \\ -d & c \end{bmatrix} \\ \begin{bmatrix} e & f \\ -f & e \end{bmatrix} & \begin{bmatrix} g & h \\ -h & g \end{bmatrix} \end{bmatrix} = \begin{bmatrix} a & b & c & d \\ -b & a & -d & c \\ e & f & g & h \\ -f & e & -h & g \end{bmatrix}$$

There is a thing of weight here. We no longer need the concept of an imaginary number.

Throughout this work, we will be concerned only with algebras that have matrix multiplication as the multiplication operation – the natural algebras. We will be concerned only with every such algebra that is an algebraic field. Thus, we will be concerned with only every natural algebraic field.

Section 1 : Chapter 2

The Nature of Algebraic Isomorphism

Two algebras over the same algebraic field[4] are said to be algebraically isomorphic if the operations of addition, multiplication, and multiplication by a scalar have the same effect upon all elements in both algebras. To be more technical, two algebras over the same field are algebraically isomorphic if:

i)   There is a bi-jective (one-to-one and onto) mapping from the elements of the first algebra to the elements of the second algebra.

ii)   The sum of any two elements within the first algebra is the element that corresponds, under the bi-jective mapping, to the sum of the corresponding two elements, under the bi-jective mapping, in the second algebra.

iii)   The product of any two elements within the first algebra is the element that corresponds, under the bi-jective mapping, to the product of the corresponding two elements, under the bi-jective mapping, in the second algebra.

iv)   The result of multiplying any element within the first algebra by a scalar is the element that corresponds, under the bi-jective mapping, to the result of multiplying the corresponding element, under the bi-jective mapping, within the second algebra by the same scalar.

Definition:

Algebraic Isomorphism

Two algebras $\{A_1, A_2\}$ over the same field, F, are isomorphic if there is a bi-jective mapping from the first algebra to the second algebra such that:

$\phi : \phi(s) \to s' \quad \forall s, t \in A_1 \quad \forall s' t' \in A_2$

i) $\phi(s+t) = s' + t'$

ii) $\phi(st) = s't'$

iii) $\phi(\alpha t) = \alpha s' \quad : \alpha \in F$

Theorem:

The algebra of matrices of the form $\{a_{11} \in \mathbb{R}, a_{12} = 0, a_{21} = 0, a_{22} = a_{11}\} \equiv \begin{bmatrix} a & 0 \\ 0 & a \end{bmatrix}$ is algebraically isomorphic to the algebra of the real numbers.

Proof:

$\phi : (\mathbb{R}, +, \times) \mapsto (M_{(2,\mathbb{R})}, +, \bullet) \quad : \quad \phi(a) \mapsto \begin{bmatrix} a & 0 \\ 0 & a \end{bmatrix}$

i) $\phi(a+e) \mapsto \begin{bmatrix} a+e & 0 \\ 0 & a+e \end{bmatrix} = \begin{bmatrix} a & 0 \\ 0 & a \end{bmatrix} + \begin{bmatrix} e & 0 \\ 0 & e \end{bmatrix}$

---

[4] Within this work, the algebraic field is always the real numbers.

ii) $\quad \phi(a \times e) \mapsto \begin{bmatrix} ae & 0 \\ 0 & ae \end{bmatrix} = \begin{bmatrix} a & 0 \\ 0 & a \end{bmatrix} \bullet \begin{bmatrix} e & 0 \\ 0 & e \end{bmatrix}$

iii) $\quad \phi(\alpha a) \mapsto \begin{bmatrix} \alpha a & 0 \\ 0 & \alpha a \end{bmatrix} = \alpha \begin{bmatrix} a & 0 \\ 0 & a \end{bmatrix}$

Although we do not give them, proofs of a similar nature apply to all (real element) matrices with all leading diagonal elements equal and all other elements zero. Hence forward, we will refer to such matrices as real number matrices.

Nomenclature:
Real Number Matrices
Square matrices with real elements with all leading diagonal elements equal and all other elements zero.[5]

Theorem:

The algebra of matrices of the form $\left\{ a_{11} \in \mathbb{R}, a_{12} \in \mathbb{R}, a_{21} = -a_{12}, a_{22} = a_{11} \right\} \equiv \begin{bmatrix} a & b \\ -b & a \end{bmatrix}$ is isomorphic to the algebra of the euclidean complex numbers.

Proof:

$$\phi : (\mathbb{C}, +, \times) \mapsto \left(M_{(2, \mathbb{R})}, +, \bullet \right) \quad : \quad \phi(a + \hat{i}b) \mapsto \begin{bmatrix} a & b \\ -b & a \end{bmatrix}$$

i) $\quad \phi\left((a + \hat{i}b) + (e + \hat{i}f) \right) \mapsto \begin{bmatrix} a+e & b+f \\ -(b+f) & a+e \end{bmatrix} = \begin{bmatrix} a & b \\ -b & a \end{bmatrix} + \begin{bmatrix} e & f \\ -f & e \end{bmatrix}$

ii) $\quad \phi\left((a + \hat{i}b) \times (e + \hat{i}f) \right) \mapsto \begin{bmatrix} ae-bf & af+be \\ -(af+be) & ae-bf \end{bmatrix} = \begin{bmatrix} a & b \\ -b & a \end{bmatrix} \bullet \begin{bmatrix} e & f \\ -f & e \end{bmatrix}$

iii) $\quad \phi\left(\alpha(a + \hat{i}b) \right) \mapsto \begin{bmatrix} \alpha a & -\alpha b \\ \alpha b & \alpha a \end{bmatrix} = \begin{bmatrix} \alpha & 0 \\ 0 & \alpha \end{bmatrix} \begin{bmatrix} a & -b \\ b & a \end{bmatrix}$

For the reader's convenience, we point out that:

$$1 \cong \begin{bmatrix} 1 & 0 \\ -0 & 1 \end{bmatrix}, \quad \hat{i} \cong \begin{bmatrix} 0 & 1 \\ -1 & 0 \end{bmatrix}, \quad a + \hat{i}b \cong \begin{bmatrix} a & b \\ -b & a \end{bmatrix}, \quad a - \hat{i}b \cong \begin{bmatrix} a & -b \\ b & a \end{bmatrix}$$

Multiplying 2×2 matrices, or larger matrices, by a real number, which is a 1×1 matrix, is an operation that is technically impossible because the matrices are of a different size. Having now proved the algebraic isomorphism of the real numbers and the $n \times n$ real number matrices, we will henceforward consider scalars to be real number matrices of the appropriate size and scalar multiplication to be included within general multiplication.

---

[5] These matrices are also known as homothetic matrices.

8

It is important to realise that the operations of addition and multiplication are the only operations that are bound to give the same results in both algebras. For example, the square-rooting operation need not give the same result in both algebras. In the 1×1 matrix form, the positive real numbers have two square roots. In 2×2 matrix form, we have:

$$\begin{bmatrix} w & x \\ y & z \end{bmatrix}\begin{bmatrix} w & x \\ y & z \end{bmatrix} = \begin{bmatrix} w^2 + xy & x(w+z) \\ y(w+z) & xy + z^2 \end{bmatrix} = \begin{bmatrix} a & 0 \\ 0 & a \end{bmatrix}$$

In the 2×2 matrix form, the real numbers have as many square roots as there are solutions of the equations:

$$w^2 + xy = a$$
$$x(w+z) = 0$$
$$y(w+z) = 0$$
$$xy + z^2 = a$$

There are an infinite number of such solutions, and hence an infinite number of square roots.

Which of the isomorphic algebras is the 'true' form of the algebra? We take the view that the true form is the one with no zeros in the matrix – the 1×1 matrices in the case of the real numbers and the 2×2 matrices in the case of the euclidean complex numbers. We refer to the larger matrices, with the unnecessary zeros, as overblown representations of the algebras.

Nomenclature:

Overblown Representation

A matrix representation of an algebra by a matrix that is larger than is necessary to include the algebraic operations[6] is called an overblown representation of that algebra.

When we seek to discover the roots of real numbers within an algebra, as we did above, because of the presence of the extra zeros, overblown representations allow more solutions than the 'true' representation of the algebra does. We avert from these overblown representations because the 'true' representations contain everything that is of the 'algebraic essence' of the algebra.

---

[6] The algebraic operations are the operations of addition, multiplication, (and, conventionally, multiplication by a scalar).

Section 1 : Chapter 3

Algebraic Field Extensions

It is conventional wisdom that the euclidean complex numbers are an algebraic field extension of the real numbers associated with the monic minimum polynomial $x^2 + 1 = 0$. We will shortly meet the hyperbolic complex numbers. These are an algebraic field and thus of the same algebraic status as the euclidean complex numbers. They contain within them the 2×2 real number matrices as a sub-algebra, as, of course, do the euclidean complex numbers. Conventional wisdom has it that, even though they contain the real numbers within them, the hyperbolic complex numbers are not an algebraic field extension of the real numbers because they are not associated with a monic minimum polynomial that will not split into linear factors over the real numbers.

We reject this conventional wisdom. Indeed, we reject the concept of algebraic extension. Instead, we adopt the concept of algebraic inclusion. We take the view that the euclidean complex number algebra contains within it a sub-algebra that is algebraically isomorphic (if overblown) to the real numbers. Similarly, the hyperbolic complex numbers contain within them a sub-algebra that is algebraically isomorphic to the real numbers. We take the view that un-splittable polynomials are nothing to do with algebraic inclusion.

We will soon meet the 3-dimensional natural algebras (the 3-dimensional complex numbers[7]). Like the lesser dimensional natural algebras, these algebras have within them a sub-algebra that is algebraically isomorphic to the real numbers. They do not include a sub-algebra that is algebraically isomorphic to any of the 2-dimensional natural algebras.

We will soon meet the 4-dimension complex numbers. Like the lesser dimensional natural algebras, these algebras have within them a sub-algebra that is algebraically isomorphic to the real numbers. They also each have a sub-algebra that is algebraically isomorphic to one of the 2-dimensional natural algebras.

This concept of algebraic inclusion rather than algebraic extension is a change to the way we view numbers. The conventional view is to begin with the natural integers and build upwards upon them. The existence of the hyperbolic complex numbers effectively topples that building. One cannot get to them by building upwards from the real numbers through unsplittable polynomials, and yet they exist. The view pressed upon us by what follows in this book is that we have 1-dimensional numbers, the real numbers; we have 2-dimensional numbers; we have 3-dimensional numbers, and so on. This is a change in the concept of number.

---

[7] A complex number algebra is the same thing as a natural algebra. We use the terms interchangeably.

Section 1 : Chapter 4

Upon Norms and Inner Products

Most algebras are created by mathematicians. It works something like this:
1) Get up in the morning and invent some objects and an addition operation that together form a linear space. We now have a linear space.
2) After breakfast, invent a multiplication operation to impose upon these objects. It cannot be any old operation; it must be closed and have one or two other properties, but, other than this, the only limit upon it is the limit of the mathematician's imagination. We now have an algebra.
3) After lunch, invent a norm to put upon this algebra. Again, it cannot be any old thing, but, within a few constraints, the only limit upon it is the limit of the mathematician's imagination. We now have a normed algebra.
4) After tea, invent an inner product operation to put upon this algebra. Again, within a few constraints, it is limited only by the mathematician's imagination. We now have a Hilbert space, and that is some algebra, and all invented by the mathematician, and all before dinner – no wonder the world so admires mathematicians.

Well, perhaps we've been a little flippant[8]. Creating an algebra is not quite that easy. The point is that the algebra, and all its bits, was invented; it was not discovered. The natural algebras are different.

Starting with the concept of linearity, writing the objects within the linear space as linear transformations (matrices), adopting the linear multiplication operation (matrix multiplication) creates the natural algebra. The natural algebra then takes control, as if it did not have control from the start. The natural algebra already has its norm within it, and the algebra imposes that norm upon the mathematician. Not only does it impose a norm upon the mathematician, but, with the norm, it also imposes a rotation matrix and a set of trigonometric functions. The natural algebra also has its inner product within it, and the algebra imposes that inner product upon the mathematician. By the same operation, the natural algebra forces one (or more) exterior product(s) on to the mathematician. Algebra creation is turned upon its head - the inmates have taken over the asylum.

---

[8] There's no 'perhaps' about it.

Section 1 : Chapter 5

Exponentiation of Matrices

The material presented in this short chapter is all standard stuff. It is presented to ensure that the reader is familiar with it.

The exponential of a matrix is:

$$\exp([A]) = \frac{1}{0!} + \frac{[A]}{1!} + \frac{[A]^2}{2!} + \frac{[A]^3}{3!} + \dots$$

However, we have:

$$\exp([A] + [B]) = \exp([A])\exp([B]) \quad \text{only if} \quad [A][B] = [B][A]$$

And, importantly:

$$\det(\exp([A])) = \exp(Trace[A]) \quad \text{so} \quad \det\left(\exp\left(\begin{bmatrix} 0 & \sim \\ \sim & 0 \end{bmatrix}\right)\right) = 1$$

12

Introduction to Section 2

In this section, we seek to familiarise the reader with the complex number algebra being done in matrix notation. We use the euclidean complex numbers to unveil an example of a central theme of this work. That central theme is that a type of space is contained within each the natural algebras – the unification of algebra and geometry.

Section 2 : Chapter 1

Euclidean Complex Numbers by Matrices

The matrix representation of a euclidean complex number, $\left(a + \hat{i}b\right)^{9}$, is:

$$\begin{bmatrix} a & b \\ -b & a \end{bmatrix}$$

where every element is a real number[10]. It is important to realize that:

$$\begin{bmatrix} a & -b \\ b & a \end{bmatrix} \quad \begin{bmatrix} -a & -b \\ b & -a \end{bmatrix} \quad \begin{bmatrix} -a & b \\ -b & -a \end{bmatrix}$$

are matrices of the same form and are also euclidean complex numbers.

Nomenclature:

Algebraic Matrix Form

If a particular form of matrix is such that matrices of that form, together with matrix multiplication, are an algebra[11], then we will refer to that form of matrix as an 'algebraic matrix form'.

Addition is, of course, matrix addition:

$$\begin{bmatrix} a & b \\ -b & a \end{bmatrix} + \begin{bmatrix} c & d \\ -d & c \end{bmatrix} = \begin{bmatrix} a+c & b+d \\ -(b+d) & a+c \end{bmatrix}$$

Multiplication is matrix multiplication:

$$\begin{bmatrix} a & b \\ -b & a \end{bmatrix}\begin{bmatrix} c & d \\ -d & c \end{bmatrix} = \begin{bmatrix} ac-bd & ad+bc \\ -(ad+bc) & ac-bd \end{bmatrix}$$

The reader should note that the form of the matrix is maintained by multiplication. The reader might like to compare the elements of the product matrix to euclidean complex number multiplication in non-matrix notation:

---

[9] $\hat{i} = \sqrt[2]{-1}$

[10] Proof of algebraic isomorphism is given in a previous chapter.

[11] This means they satisfy the algebra axioms – see appendices.

$$\left(a+\hat{i}b\right)\left(c+\hat{i}d\right)=\left(ac-bd\right)+\hat{i}\left(ad+bc\right)$$

Because the euclidean complex numbers are 2×2 matrices, they each have a determinant.

$$\det\left(\begin{bmatrix} a & b \\ -b & a \end{bmatrix}\right)=a^2+b^2$$

The reader should note that the determinant of a euclidean complex number is the same as the modulus (*norm*) of the euclidean complex number $\left(a+\hat{i}b\right)$.

The inverse of a euclidean complex number is:

$$\begin{bmatrix} a & b \\ -b & a \end{bmatrix}^{-1}=\frac{1}{a^2+b^2}\begin{bmatrix} a & -b \\ b & a \end{bmatrix}$$

The conjugate of a euclidean complex number in matrix form is the adjoint matrix of that complex number. The adjoint of a matrix is the inverse matrix multiplied by the determinant of that matrix:

$$Conjugate\left(\begin{bmatrix} a & b \\ -b & a \end{bmatrix}\right)=\left(a^2+b^2\right)\begin{bmatrix} a & b \\ -b & a \end{bmatrix}^{-1}=\frac{a^2+b^2}{a^2+b^2}\begin{bmatrix} a & -b \\ b & a \end{bmatrix}=\begin{bmatrix} a & -b \\ b & a \end{bmatrix}$$

A Word of Warning:
   Some texts say that the conjugate of a euclidean complex number written in matrix notation is the transpose of that matrix. It is not so; the conjugate is the adjoint of that matrix, and it is accidental that, in the case of euclidean complex numbers, the adjoint and the transpose are the same. This distinction is of no matter for euclidean complex numbers but of great matter for the other algebras that we shall meet.

Division is by use of the conjugate, of course:

$$\frac{\begin{bmatrix} a & b \\ -b & a \end{bmatrix}}{\begin{bmatrix} c & d \\ -d & c \end{bmatrix}}=\frac{\begin{bmatrix} a & b \\ -b & a \end{bmatrix}\begin{bmatrix} c & -d \\ d & c \end{bmatrix}}{\begin{bmatrix} c & d \\ -d & c \end{bmatrix}\begin{bmatrix} c & -d \\ d & c \end{bmatrix}}=\frac{\begin{bmatrix} a & b \\ -b & a \end{bmatrix}\begin{bmatrix} c & -d \\ d & c \end{bmatrix}}{\begin{bmatrix} c^2+d^2 & 0 \\ 0 & c^2+d^2 \end{bmatrix}}$$

$$=\begin{bmatrix} \dfrac{1}{c^2+d^2} & 0 \\ 0 & \dfrac{1}{c^2+d^2} \end{bmatrix}\begin{bmatrix} a & b \\ -b & a \end{bmatrix}\begin{bmatrix} c & -d \\ d & c \end{bmatrix}$$

which one can compare to:

$$\frac{a+\hat{i}b}{c+\hat{i}d} = \frac{\left(a+\hat{i}b\right)\left(c-\hat{i}d\right)}{\left(c+\hat{i}d\right)\left(c-\hat{i}d\right)} = \frac{\left(a+\hat{i}b\right)\left(c-\hat{i}d\right)}{c^2+d^2}$$

We have:

$$\begin{bmatrix} a & b \\ -b & a \end{bmatrix} = \begin{bmatrix} a & 0 \\ 0 & a \end{bmatrix} + \begin{bmatrix} 0 & b \\ -b & 0 \end{bmatrix} = \begin{bmatrix} a & 0 \\ 0 & a \end{bmatrix}\begin{bmatrix} 1 & 0 \\ 0 & 1 \end{bmatrix} + \begin{bmatrix} b & 0 \\ 0 & b \end{bmatrix}\begin{bmatrix} 0 & 1 \\ -1 & 0 \end{bmatrix}$$

and the reader might like to think of the diagonal matrix (the one with the a's in it) as the real part of the euclidean complex number and the other matrix (the one with the b's in it) as the imaginary part. As previously shown, matrices of the form:

$$\begin{bmatrix} a & 0 \\ 0 & a \end{bmatrix}$$

are algebraically isomorphic to the real numbers. The reader should note that we have no need of the imaginary number $\hat{i} = \sqrt{-1}$ (all the elements of the matrix are real numbers), but that we do have need of the matrix:

$$\begin{bmatrix} 0 & 1 \\ -1 & 0 \end{bmatrix}^2 = \begin{bmatrix} -1 & 0 \\ 0 & -1 \end{bmatrix}$$

Section 2 : Chapter 2

Matrix Differentiation and the Cauchy Riemann Equations

A function of a 2×2 matrix is a mapping from a 2×2 matrix to a 2×2 matrix of the form:

$$f\left(\begin{bmatrix} a & b \\ c & d \end{bmatrix}\right) \rightarrow \begin{bmatrix} u(a,b,c,d) & v(a,b,c,d) \\ s(a,b,c,d) & t(a,b,c,d) \end{bmatrix}$$

In this chapter, we consider only matrices of the form:

$$f\left(\begin{bmatrix} a & b \\ -b & a \end{bmatrix}\right) \rightarrow \begin{bmatrix} u(a,b) & v(a,b) \\ -v(a,b) & u(a,b) \end{bmatrix}$$

Matrix Differentiation:

In general, we do not know how to differentiate matrices of variables, but we do know how to differentiate real number variables. Because we know how to differentiate real number variables, we know how to differentiate matrix variables of the form:

$$\begin{bmatrix} u(a,b) & 0 \\ 0 & u(a,b) \end{bmatrix}$$

We approach the differentiation of matrices of the form:

$$\begin{bmatrix} u(a,b) & v(a,b) \\ -v(a,b) & u(a,b) \end{bmatrix}$$

by splitting them up and doing a bit of matrix arithmetic.

$$\begin{bmatrix} u(a,b) & v(a,b) \\ -v(a,b) & u(a,b) \end{bmatrix} = \begin{bmatrix} u(a,b) & 0 \\ 0 & u(a,b) \end{bmatrix} + \begin{bmatrix} 0 & v(a,b) \\ -v(a,b) & 0 \end{bmatrix}$$

$$= \begin{bmatrix} u(a,b) & 0 \\ 0 & u(a,b) \end{bmatrix} + \begin{bmatrix} 0 & 1 \\ -1 & 0 \end{bmatrix}\begin{bmatrix} v(a,b) & 0 \\ 0 & v(a,b) \end{bmatrix}$$

This is the sum of two real number variables where the matrix $\begin{bmatrix} 0 & 1 \\ -1 & 0 \end{bmatrix}$ is a constant.

Technique:

Matrix Differentiation

You take a matrix. You split it into pieces. You turn each piece into a real number matrix and a constant. Then you can differentiate it. (Clearly, this will work with only 'special' types of matrix.)

We will be differentiating with respect to a real number variable. This means that we differentiate with respect to either: $\begin{bmatrix} a & 0 \\ 0 & a \end{bmatrix}$ or $\begin{bmatrix} 0 & 1 \\ -1 & 0 \end{bmatrix}\begin{bmatrix} b & 0 \\ 0 & b \end{bmatrix}$

We consider functions of euclidean complex numbers. These are of the form:

$$f\left(\begin{bmatrix} a & b \\ -b & a \end{bmatrix}\right) \rightarrow \begin{bmatrix} u(a,b) & v(a,b) \\ -v(a,b) & u(a,b) \end{bmatrix}$$

Differentiation of the function $f\left(\begin{bmatrix} a & b \\ -b & a \end{bmatrix}\right)$ with respect to $\begin{bmatrix} a & 0 \\ 0 & a \end{bmatrix}$ at point $\begin{bmatrix} \alpha_{11} & -\alpha_{21} \\ \alpha_{21} & \alpha_{11} \end{bmatrix}$

is (holding b constant):

$$f'\left(\begin{bmatrix} \alpha_{11} & \alpha_{12} \\ -\alpha_{21} & \alpha_{11} \end{bmatrix}\right) = \lim_{\begin{bmatrix} a & \alpha_{21} \\ -\alpha_{21} & a \end{bmatrix} \rightarrow \begin{bmatrix} \alpha_{11} & \alpha_{21} \\ -\alpha_{21} & \alpha_{11} \end{bmatrix}} \frac{\begin{bmatrix} u(a,\alpha_{21}) & v(a,\alpha_{21}) \\ -v(a,\alpha_{21}) & u(a,\alpha_{21}) \end{bmatrix} - \begin{bmatrix} u(\alpha_{11},\alpha_{21}) & v(\alpha_{11},\alpha_{21}) \\ -v(\alpha_{11},\alpha_{21}) & u(\alpha_{11},\alpha_{21}) \end{bmatrix}}{\begin{bmatrix} a & \alpha_{21} \\ -\alpha_{21} & a \end{bmatrix} - \begin{bmatrix} \alpha_{11} & \alpha_{21} \\ -\alpha_{21} & \alpha_{11} \end{bmatrix}}$$

$$f'\left(\begin{bmatrix} \alpha_{11} & \alpha_{12} \\ -\alpha_{21} & \alpha_{11} \end{bmatrix}\right) = \lim_{\begin{bmatrix} a & \alpha_{21} \\ -\alpha_{21} & a \end{bmatrix} \rightarrow \begin{bmatrix} \alpha_{11} & \alpha_{21} \\ -\alpha_{21} & \alpha_{11} \end{bmatrix}} \frac{\begin{bmatrix} u(a,\alpha_{21}) & 0 \\ 0 & u(a,\alpha_{21}) \end{bmatrix} - \begin{bmatrix} u(\alpha_{11},\alpha_{21}) & 0 \\ 0 & u(\alpha_{11},\alpha_{21}) \end{bmatrix}}{\begin{bmatrix} a & 0 \\ 0 & a \end{bmatrix} - \begin{bmatrix} \alpha_{11} & 0 \\ 0 & \alpha_{11} \end{bmatrix}}$$

$$+ \lim_{\begin{bmatrix} a & \alpha_{21} \\ -\alpha_{21} & a \end{bmatrix} \rightarrow \begin{bmatrix} \alpha_{11} & \alpha_{21} \\ -\alpha_{21} & \alpha_{11} \end{bmatrix}} \frac{\begin{bmatrix} 0 & v(a,\alpha_{21}) \\ -v(a,\alpha_{21}) & 0 \end{bmatrix} - \begin{bmatrix} 0 & v(\alpha_{11},\alpha_{21}) \\ -v(\alpha_{11},\alpha_{21}) & 0 \end{bmatrix}}{\begin{bmatrix} a & 0 \\ 0 & a \end{bmatrix} - \begin{bmatrix} \alpha_{11} & 0 \\ 0 & \alpha_{11} \end{bmatrix}}$$

$$f'\left(\begin{bmatrix} \alpha_{11} & \alpha_{12} \\ -\alpha_{21} & \alpha_{11} \end{bmatrix}\right) = \lim_{\begin{bmatrix} a & \alpha_{21} \\ -\alpha_{21} & a \end{bmatrix} \rightarrow \begin{bmatrix} \alpha_{11} & \alpha_{21} \\ -\alpha_{21} & \alpha_{11} \end{bmatrix}} \frac{\begin{bmatrix} u(a,\alpha_{21}) & 0 \\ 0 & u(a,\alpha_{21}) \end{bmatrix} - \begin{bmatrix} u(\alpha_{11},\alpha_{21}) & 0 \\ 0 & u(\alpha_{11},\alpha_{21}) \end{bmatrix}}{\begin{bmatrix} a & 0 \\ 0 & a \end{bmatrix} - \begin{bmatrix} \alpha_{11} & 0 \\ 0 & \alpha_{11} \end{bmatrix}}$$

$$+ \begin{bmatrix} 0 & 1 \\ -1 & 0 \end{bmatrix} \lim_{\begin{bmatrix} a & \alpha_{21} \\ -\alpha_{21} & a \end{bmatrix} \rightarrow \begin{bmatrix} \alpha_{11} & \alpha_{21} \\ -\alpha_{21} & \alpha_{11} \end{bmatrix}} \frac{\begin{bmatrix} v(a,\alpha_{21}) & 0 \\ 0 & v(a,\alpha_{21}) \end{bmatrix} - \begin{bmatrix} v(\alpha_{11},\alpha_{21}) & 0 \\ 0 & v(\alpha_{11},\alpha_{21}) \end{bmatrix}}{\begin{bmatrix} a & 0 \\ 0 & a \end{bmatrix} - \begin{bmatrix} \alpha_{11} & 0 \\ 0 & \alpha_{11} \end{bmatrix}}$$

$$f'\left(\begin{bmatrix} \alpha_{11} & \alpha_{12} \\ -\alpha_{21} & \alpha_{11} \end{bmatrix}\right) = \begin{bmatrix} \dfrac{\partial u}{\partial a} & 0 \\ 0 & \dfrac{\partial u}{\partial a} \end{bmatrix} + \begin{bmatrix} 0 & 1 \\ -1 & 0 \end{bmatrix} \begin{bmatrix} \dfrac{\partial v}{\partial a} & 0 \\ 0 & \dfrac{\partial v}{\partial a} \end{bmatrix}$$

Which in non-matrix notation would be written as:

$$f'(\alpha_{11}, \alpha_{21}) = \frac{\partial u}{\partial a}(\alpha_{11}, \alpha_{21}) + \hat{i}\frac{\partial v}{\partial a}(\alpha_{11}, \alpha_{21})$$

We write it as:

$$f'\left(\begin{bmatrix} \alpha_{11} & \alpha_{12} \\ -\alpha_{21} & \alpha_{11} \end{bmatrix}\right) = \begin{bmatrix} \dfrac{\partial u}{\partial a}(\alpha_{11}, \alpha_{21}) & \dfrac{\partial v}{\partial a}(\alpha_{11}, \alpha_{21}) \\ -\dfrac{\partial v}{\partial a}(\alpha_{11}, \alpha_{21}) & \dfrac{\partial u}{\partial a}(\alpha_{11}, \alpha_{21}) \end{bmatrix}$$

We now repeat the procedure holding a constant.

Differentiation of the function $f\left(\begin{bmatrix} a & b \\ -b & a \end{bmatrix}\right)$ with respect to $\begin{bmatrix} 0 & 1 \\ -1 & 0 \end{bmatrix}\begin{bmatrix} b & 0 \\ 0 & b \end{bmatrix} = \begin{bmatrix} 0 & b \\ -b & 0 \end{bmatrix}$ at

point $\begin{bmatrix} \alpha_{11} & \alpha_{21} \\ -\alpha_{21} & \alpha_{11} \end{bmatrix}$ is (holding a constant):

$$f'\left(\begin{bmatrix} \alpha_{11} & \alpha_{12} \\ -\alpha_{21} & \alpha_{11} \end{bmatrix}\right) = \lim_{\left[\begin{smallmatrix} \alpha_{11} & b \\ -b & \alpha_{11} \end{smallmatrix}\right] \to \left[\begin{smallmatrix} \alpha_{11} & \alpha_{21} \\ -\alpha_{21} & \alpha_{11} \end{smallmatrix}\right]} \frac{\begin{bmatrix} u(\alpha_{11},b) & v(\alpha_{11},b) \\ -v(\alpha_{11},b) & u(\alpha_{11},b) \end{bmatrix} - \begin{bmatrix} u(\alpha_{11},\alpha_{21}) & v(\alpha_{11},\alpha_{21}) \\ -v(\alpha_{11},\alpha_{21}) & u(\alpha_{11},\alpha_{21}) \end{bmatrix}}{\begin{bmatrix} \alpha_{11} & b \\ -b & \alpha_{11} \end{bmatrix} - \begin{bmatrix} \alpha_{11} & \alpha_{21} \\ -\alpha_{21} & \alpha_{11} \end{bmatrix}}$$

$$f'\left(\begin{bmatrix} \alpha_{11} & \alpha_{12} \\ -\alpha_{21} & \alpha_{11} \end{bmatrix}\right) = \lim_{\left[\begin{smallmatrix} \alpha_{11} & b \\ -b & \alpha_{11} \end{smallmatrix}\right] \to \left[\begin{smallmatrix} \alpha_{11} & \alpha_{21} \\ -\alpha_{21} & \alpha_{11} \end{smallmatrix}\right]} \frac{\begin{bmatrix} u(\alpha_{11},b) & 0 \\ 0 & u(\alpha_{11},b) \end{bmatrix} - \begin{bmatrix} u(\alpha_{11},\alpha_{21}) & 0 \\ 0 & u(\alpha_{11},\alpha_{21}) \end{bmatrix}}{\begin{bmatrix} 0 & b \\ -b & 0 \end{bmatrix} - \begin{bmatrix} 0 & \alpha_{21} \\ -\alpha_{21} & 0 \end{bmatrix}}$$

$$+ \lim_{\left[\begin{smallmatrix} \alpha_{11} & b \\ -b & \alpha_{11} \end{smallmatrix}\right] \to \left[\begin{smallmatrix} \alpha_{11} & \alpha_{21} \\ -\alpha_{21} & \alpha_{11} \end{smallmatrix}\right]} \frac{\begin{bmatrix} 0 & v(\alpha_{11},b) \\ -v(\alpha_{11},b) & 0 \end{bmatrix} - \begin{bmatrix} 0 & v(\alpha_{11},\alpha_{21}) \\ -v(\alpha_{11},\alpha_{21}) & 0 \end{bmatrix}}{\begin{bmatrix} 0 & b \\ -b & 0 \end{bmatrix} - \begin{bmatrix} 0 & \alpha_{21} \\ -\alpha_{21} & 0 \end{bmatrix}}$$

We now change the variable that we are differentiating with respect to from $\begin{bmatrix} 0 & b \\ -b & 0 \end{bmatrix}$ to

$\begin{bmatrix} 0 & 1 \\ -1 & 0 \end{bmatrix}\begin{bmatrix} b & 0 \\ 0 & b \end{bmatrix}$.

$$f'\left(\begin{bmatrix} \alpha_{11} & \alpha_{12} \\ -\alpha_{21} & \alpha_{11} \end{bmatrix}\right) = \lim_{\begin{bmatrix} \alpha_{11} & b \\ -b & \alpha_{11} \end{bmatrix} \to \begin{bmatrix} \alpha_{11} & \alpha_{21} \\ -\alpha_{21} & \alpha_{11} \end{bmatrix}} \frac{\begin{bmatrix} u(\alpha_{11},b) & 0 \\ 0 & u(\alpha_{11},b) \end{bmatrix} - \begin{bmatrix} u(\alpha_{11},\alpha_{21}) & 0 \\ 0 & u(\alpha_{11},\alpha_{21}) \end{bmatrix}}{\begin{bmatrix} 0 & 1 \\ -1 & 0 \end{bmatrix}\left(\begin{bmatrix} b & 0 \\ 0 & b \end{bmatrix} - \begin{bmatrix} \alpha_{21} & 0 \\ 0 & \alpha_{21} \end{bmatrix}\right)}$$

$$+ \begin{bmatrix} 0 & 1 \\ -1 & 0 \end{bmatrix} \lim_{\begin{bmatrix} \alpha_{11} & b \\ -b & \alpha_{11} \end{bmatrix} \to \begin{bmatrix} \alpha_{11} & \alpha_{21} \\ -\alpha_{21} & \alpha_{11} \end{bmatrix}} \frac{\begin{bmatrix} v(\alpha_{11},b) & 0 \\ 0 & v(\alpha_{11},b) \end{bmatrix} - \begin{bmatrix} v(\alpha_{11},\alpha_{21}) & 0 \\ 0 & v(\alpha_{11},\alpha_{21}) \end{bmatrix}}{\begin{bmatrix} 0 & 1 \\ -1 & 0 \end{bmatrix}\left(\begin{bmatrix} b & 0 \\ 0 & b \end{bmatrix} - \begin{bmatrix} \alpha_{21} & 0 \\ 0 & \alpha_{21} \end{bmatrix}\right)}$$

$$f'\left(\begin{bmatrix} \alpha_{11} & \alpha_{12} \\ -\alpha_{21} & \alpha_{11} \end{bmatrix}\right) = \frac{1}{\begin{bmatrix} 0 & 1 \\ -1 & 0 \end{bmatrix}} \begin{bmatrix} \frac{\partial u}{\partial b} & 0 \\ 0 & \frac{\partial u}{\partial b} \end{bmatrix} + \begin{bmatrix} \frac{\partial v}{\partial b} & 0 \\ 0 & \frac{\partial v}{\partial b} \end{bmatrix}$$

$$= \begin{bmatrix} 0 & -1 \\ 1 & 0 \end{bmatrix} \begin{bmatrix} \frac{\partial u}{\partial b} & 0 \\ 0 & \frac{\partial u}{\partial b} \end{bmatrix} + \begin{bmatrix} \frac{\partial v}{\partial b} & 0 \\ 0 & \frac{\partial v}{\partial b} \end{bmatrix}$$

Which in non-matrix notation would be written as:

$$f'(\alpha_{11},\alpha_{21}) = \frac{\partial v}{\partial b}(\alpha_{11},\alpha_{21}) - \hat{i}\frac{\partial u}{\partial b}(\alpha_{11},\alpha_{21})$$

We write it as:

$$f'\left(\begin{bmatrix} \alpha_{11} & \alpha_{12} \\ -\alpha_{21} & \alpha_{11} \end{bmatrix}\right) = \begin{bmatrix} \frac{\partial v}{\partial b}(\alpha_{11},\alpha_{21}) & -\frac{\partial u}{\partial b}(\alpha_{11},\alpha_{21}) \\ \frac{\partial u}{\partial b}(\alpha_{11},\alpha_{21}) & \frac{\partial v}{\partial b}(\alpha_{11},\alpha_{21}) \end{bmatrix}$$

The euclidean complex matrix function is differentiable only if the two differentiations are equal, ie:

$$\frac{\partial u}{\partial a} = \frac{\partial v}{\partial b} \quad \& \quad \frac{\partial u}{\partial b} = -\frac{\partial v}{\partial a}$$

These are the familiar Cauchy-Riemann equations of euclidean complex analysis.

The Cauchy Riemann equations say that we will get the same answer regardless of which variable we differentiate with respect to. This means that the space (Argand diagram) that we are working in is isotropic. It is the (subjective) insistence upon the isotropy of this space that leads to the Cauchy Riemann equations, and so we are going around in circles.

Summary: Differentiate by:

1) Splitting the function matrix into the different functions $\{u(a,b), v(a,b)\}$:

$$\begin{bmatrix} u(a,b) & v(a,b) \\ -v(a,b) & u(a,b) \end{bmatrix} = \begin{bmatrix} u(a,b) & 0 \\ 0 & u(a,b) \end{bmatrix} + \begin{bmatrix} 0 & v(a,b) \\ -v(a,b) & 0 \end{bmatrix}$$

2) Factor this expression to get all functions on the leading diagonal:

$$\begin{bmatrix} u(a,b) & 0 \\ 0 & u(a,b) \end{bmatrix} + \begin{bmatrix} 0 & v(a,b) \\ -v(a,b) & 0 \end{bmatrix} =$$

$$\begin{bmatrix} 1 & 0 \\ 0 & 1 \end{bmatrix}\begin{bmatrix} u(a,b) & 0 \\ 0 & u(a,b) \end{bmatrix} + \begin{bmatrix} 0 & 1 \\ -1 & 0 \end{bmatrix}\begin{bmatrix} v(a,b) & 0 \\ 0 & v(a,b) \end{bmatrix} :$$

3) We are differentiating with respect to a variable matrix. In this case either:

$$\begin{bmatrix} a & 0 \\ 0 & a \end{bmatrix} \quad \text{or} \quad \begin{bmatrix} 0 & b \\ -b & 0 \end{bmatrix}$$

Factor these matrices to get all variables on the leading diagonal:

$$\begin{bmatrix} 1 & 0 \\ 0 & 1 \end{bmatrix}\begin{bmatrix} a & 0 \\ 0 & a \end{bmatrix} \quad \text{or} \quad \begin{bmatrix} 0 & 1 \\ -1 & 0 \end{bmatrix}\begin{bmatrix} b & 0 \\ 0 & b \end{bmatrix}$$

4) Choose which variable matrix to differentiate with respect to, differentiate the functions in the function matrices by that variable, and divide the answer by the factor attached to the variable matrix:

$$\frac{\partial\left(\begin{bmatrix} u(a,b) & v(a,b) \\ -v(a,b) & u(a,b) \end{bmatrix} \right)}{\partial a} = \frac{\begin{bmatrix} 1 & 0 \\ 0 & 1 \end{bmatrix}\begin{bmatrix} \frac{\partial u}{\partial a} & 0 \\ 0 & \frac{\partial u}{\partial a} \end{bmatrix} + \begin{bmatrix} 0 & 1 \\ -1 & 0 \end{bmatrix}\begin{bmatrix} \frac{\partial v}{\partial a} & 0 \\ 0 & \frac{\partial v}{\partial a} \end{bmatrix}}{} = \begin{bmatrix} \frac{\partial u}{\partial a} & \frac{\partial v}{\partial a} \\ -\frac{\partial v}{\partial a} & \frac{\partial u}{\partial a} \end{bmatrix}$$

$$\frac{\partial\left(\begin{bmatrix} u(a,b) & v(a,b) \\ -v(a,b) & u(a,b) \end{bmatrix} \right)}{\partial b} = \frac{\begin{bmatrix} 1 & 0 \\ 0 & 1 \end{bmatrix}\begin{bmatrix} \frac{\partial u}{\partial b} & 0 \\ 0 & \frac{\partial u}{\partial b} \end{bmatrix} + \begin{bmatrix} 0 & 1 \\ -1 & 0 \end{bmatrix}\begin{bmatrix} \frac{\partial v}{\partial b} & 0 \\ 0 & \frac{\partial v}{\partial b} \end{bmatrix}}{} = \begin{bmatrix} \frac{\partial v}{\partial b} & -\frac{\partial u}{\partial b} \\ \frac{\partial u}{\partial b} & \frac{\partial v}{\partial b} \end{bmatrix}$$

All of the above just moves the positions of the differentials about, but watch the minus signs.

Section 2 : Chapter 3

The Polar Form of Euclidean Complex Numbers

We have:

$$\begin{bmatrix} a & 0 \\ 0 & a \end{bmatrix}\begin{bmatrix} 0 & b \\ -b & 0 \end{bmatrix} = \begin{bmatrix} 0 & b \\ -b & 0 \end{bmatrix}\begin{bmatrix} a & 0 \\ 0 & a \end{bmatrix}$$

Because these sub-matrices are multiplicatively commutative, we can exponentiate them and split them.

$$\exp\left(\begin{bmatrix} a & b \\ -b & a \end{bmatrix}\right) = \exp\left(\begin{bmatrix} a & 0 \\ 0 & a \end{bmatrix} + \begin{bmatrix} 0 & b \\ -b & 0 \end{bmatrix}\right)$$

$$= \exp\left(\begin{bmatrix} a & 0 \\ 0 & a \end{bmatrix}\right)\exp\left(\begin{bmatrix} 0 & b \\ -b & 0 \end{bmatrix}\right)$$

$$= \begin{bmatrix} e^a & 0 \\ 0 & e^a \end{bmatrix}\begin{bmatrix} 1 - \dfrac{b^2}{2!} + \dfrac{b^4}{4!} - \dots & b - \dfrac{b^3}{3!} + \dfrac{b^5}{5!} - \dots \\ -b + \dfrac{b^3}{3!} - \dfrac{b^5}{5!} + \dots & 1 - \dfrac{b^2}{2!} + \dfrac{b^4}{4!} - \dots \end{bmatrix}$$

$$= \begin{bmatrix} r & 0 \\ 0 & r \end{bmatrix}\begin{bmatrix} \cos b & \sin b \\ -\sin b & \cos b \end{bmatrix}$$

The algebra of the real numbers is associated with the 1-dimensional space that is the real number line. Thus, the real number matrix is a length matrix. The matrix with the trigonometric functions in it is the rotation matrix of 2-dimensional euclidean space. Because the b sub-matrix has trace equal to zero, we have:

$$\det\left(\exp\left(\begin{bmatrix} 0 & b \\ -b & 0 \end{bmatrix}\right)\right) = \exp(0) = 1$$

and this is the origin of the identity:

$$\cos^2 b + \sin^2 b = 1$$

Because the complex numbers are closed under exponentiation[12], we have:

$$\begin{bmatrix} x & y \\ -y & x \end{bmatrix} = \begin{bmatrix} r & 0 \\ 0 & r \end{bmatrix}\begin{bmatrix} \cos b & \sin b \\ -\sin b & \cos b \end{bmatrix}$$

---

[12] Exponentiation is no more than multiplication and addition.

$$\det\left(\begin{bmatrix} x & y \\ -y & x \end{bmatrix}\right) = \det\left(\begin{bmatrix} r & 0 \\ 0 & r \end{bmatrix}\begin{bmatrix} \cos b & \sin b \\ -\sin b & \cos b \end{bmatrix}\right)$$

$$x^2 + y^2 = r^2$$

Leading to:

$$r = \sqrt{x^2 + y^2}$$

which is the metric of 2-dimensional euclidean space. Hence the 'euclidean' in the name 'euclidean complex numbers'.

The determinant of a 2×2 matrix is, by definition, the area of the parallelogram formed by the two vectors that are the columns (or rows) of the matrix. Thus, we ought not to be surprised that the square root of the determinant is a distance. Within this work, we will find that, in general, the distance function of an n-dimensional natural space is the nth root of the determinant of the algebraic matrix form associated with that space[13].

We have:

$$\begin{bmatrix} x & y \\ -y & x \end{bmatrix} = \begin{bmatrix} r & 0 \\ 0 & r \end{bmatrix}\begin{bmatrix} \cos b & \sin b \\ -\sin b & \cos b \end{bmatrix}$$

$$\begin{bmatrix} \dfrac{x}{r} & \dfrac{y}{r} \\ -\dfrac{y}{r} & \dfrac{x}{r} \end{bmatrix} = \begin{bmatrix} \cos b & \sin b \\ -\sin b & \cos b \end{bmatrix}$$

The determinant of the rotation matrix is unity, and so the determinant of the $\left(\dfrac{x}{r}, \dfrac{y}{r}\right)$ matrix is also unity. We have normalised the $\{x, y\}$ matrix. We see that the polar form of a normalised matrix is such that the trigonometric functions are equal to the normalised values upon the axes:

$$\cos b = \frac{x}{r}$$

$$\sin b = \frac{y}{r}$$

In other words, we have it that the trigonometric functions are projections from the normalised position (x, y) on to the axes of the space. We knew this anyway. The cosine function is the projection from the unit circle on to the horizontal axis, and the sine function is the projection from the unit circle on to the vertical axis. However, if, as we will, we meet unfamiliar functions in place of the sine and cosine, the same procedure will confirm them to be projections from the unit 'sphere'[14] on to the different axes.

There are three points to be made:

---

[13] Of course, the determinant of a matrix is invariant under change of basis.

[14] We use the word sphere in its general multi-dimensional sense that includes circles, 4-dimensional spheres, etc.

1) The trigonometric functions associated with 2-dimensional euclidean space fall out of the algebra of the euclidean complex numbers and declare themselves to be projections from the unit circle on to the axes of the space.

2) The rotation matrix of 2-dimensional euclidean space falls out of the algebra of the euclidean complex numbers.

3) The metric of 2-dimensional euclidean space falls out of the algebra of the euclidean complex numbers and declares itself to be the distance function of this space. We take the view that this is sufficient to say that 2-dimensional euclidean space is contained within the algebra of the euclidean complex numbers[15]. Worth a theorem, methinks.

Theorem:
2-dimensional euclidean space is contained within the algebra of the euclidean complex numbers.
Proof:
See above.

With this theorem, we have effectively unified the geometry of 2-dimensional euclidean space and the algebra of the euclidean complex numbers.

If we can show that euclidean complex numbers exist even if the universe does not exist[16], then we can show that space has no choice other than to exist.[17]

We point out that the distance function (the metric) is symmetric in $\{x, y\}$ by which we mean that, if $\{x, y\}$ are swapped around, the output of the distance function is unaffected. This means that 2-dimensional euclidean space is isotropic[18]. In particular, we point out that the x-axis is of the same nature as the y-axis; if one is a spatial axis, then so is the other; if one is a temporal axis, then so is the other.

Thus, the space that has fallen out of the euclidean complex numbers is isotropic. This leads to the necessity of the Cauchy Riemann equations.

Important Point:
What we have done above will be repeated for other algebraic matrix forms. In each case, we will get functions that are projections from the unit sphere on to the axes. In each case, we will get a rotation matrix. In each case, we will get a distance function.

---

[15] This is a view and not a mathematical deduction. The view depends upon just what one requires to make a space. Many mathematicians are willing to accept the metric only.
[16] Which we shall not be attempting to do.
[17] In the case of the hyperbolic complex numbers that we are yet to meet, we get space-time within the algebra.
[18] It is also homogeneous because homogeneity is a property of the real numbers which sit upon the axes.

Section 2 : Chapter 4

Vectors and Inner Products

To be a vector, a mathematical object has to be one of a set of similar objects that satisfy the axioms of a linear space. Indeed, linear spaces are often called vector spaces. An algebraic field is a linear space together with a suitable multiplication operation, and so it ought to be no surprise that algebraic fields can do everything that vectors can do.

Within physics, we often find the use of mathematical, arrow-like, objects called vectors (usually written in boldface type or with a line underneath them) to describe various physical situations. These mathematical, arrow-like, objects, together with two operations called the inner product (also known as dot product or scalar product) and the cross product are referred to as 'vector algebra'. This is a terminological inexactitude.

This set of arrow-like objects with these two operations is not an algebra, and neither of the two operations is a bona-fide algebraic operation. From a mathematical point of view, so-called 'vector algebra' is a mess. The only reason physicists like vector algebra is because it works perfectly[19]. To make vector algebra into a proper algebra, we need an algebraic vector product. An algebraic vector product is an appropriate[20] way of multiplying two vectors together to produce another vector.

There is one-to-one correspondence between the 2-dimensional vectors and the euclidean complex number matrices.

$$\begin{bmatrix} a & b \\ -b & a \end{bmatrix} \leftrightarrow \begin{bmatrix} a \\ b \end{bmatrix}$$

The euclidean complex number matrix, as well as being a euclidean complex number, is a position vector in 2-dimensional euclidean space. Within the complex number algebra, there is the product of two complex numbers, which is a complex number, of course. This is our algebraic vector product. The algebraic way to do vector algebra is to do it using square matrices that form an algebraic field[21].

Within traditional vector algebra, we have the inner product and the cross product. There is such an operation within the euclidean complex numbers. Within the euclidean complex numbers, this operation is no more than a means of calculating the angle subtended at the origin between two vectors. Take two euclidean complex numbers in both cartesian and polar forms:

$$\left\{ \begin{bmatrix} x & y \\ -y & x \end{bmatrix}, \begin{bmatrix} t & z \\ -z & t \end{bmatrix} \right\} :: \left\{ \begin{bmatrix} r & 0 \\ 0 & r \end{bmatrix} \begin{bmatrix} \cos\chi & \sin\chi \\ -\sin\chi & \cos\chi \end{bmatrix}, \begin{bmatrix} s & 0 \\ 0 & s \end{bmatrix} \begin{bmatrix} \cos(\chi+\phi) & \sin(\chi+\phi) \\ -\sin(\chi+\phi) & \cos(\chi+\phi) \end{bmatrix} \right\}$$

where ϕ is the angle between the two complex numbers subtended at the origin. Normalize the matrices; take the conjugate of either one of the pair:

---

[19] And, for this price, they sell their souls - bunch of tarts.

[20] It is appropriate if it satisfies the algebraic field axioms.

[21] If we attempt to do it with general matrices, then, because they do not commute, we cannot form a polar form. If we cannot form a polar form, we do not have a geometric space.

$$\begin{bmatrix} \dfrac{x}{\sqrt{x^2+y^2}} & \dfrac{-y}{\sqrt{x^2+y^2}} \\[4mm] \dfrac{y}{\sqrt{x^2+y^2}} & \dfrac{x}{\sqrt{x^2+y^2}} \end{bmatrix} \begin{bmatrix} \dfrac{t}{\sqrt{t^2+z^2}} & \dfrac{z}{\sqrt{t^2+z^2}} \\[4mm] \dfrac{-z}{\sqrt{t^2+z^2}} & \dfrac{t}{\sqrt{t^2+z^2}} \end{bmatrix}$$

$$\begin{bmatrix} \cos\chi & -\sin\chi \\ \sin\chi & \cos\chi \end{bmatrix} \begin{bmatrix} \cos(\chi+\phi) & \sin(\chi+\phi) \\ -\sin(\chi+\phi) & \cos(\chi+\phi) \end{bmatrix}$$

Multiply the resulting pair together leading to:

$$\begin{bmatrix} \dfrac{xt+yz}{\sqrt{x^2+y^2}\sqrt{t^2+z^2}} & \dfrac{xz-yt}{\sqrt{x^2+y^2}\sqrt{t^2+z^2}} \\[5mm] -\left(\dfrac{xz-yt}{\sqrt{x^2+y^2}\sqrt{t^2+z^2}}\right) & \dfrac{xt+yz}{\sqrt{x^2+y^2}\sqrt{t^2+z^2}} \end{bmatrix} : \begin{bmatrix} \cos\phi & \sin\phi \\ -\sin\phi & \cos\phi \end{bmatrix}$$

Equating the elements gives the vector inner product and the exterior product (sloppily, the cross product without direction)[22].

$$\cos\phi = \frac{xt+yz}{\sqrt{x^2+y^2}\sqrt{t^2+z^2}}$$

$$\sin\phi = \frac{xz-yt}{\sqrt{x^2+y^2}\sqrt{t^2+z^2}}$$

Note:
The expression in 2-dimensional euclidean vector algebra for the angle between two vectors, $\{\vec{a}=[x,y],\vec{b}=[t,z]\}$ is:

$$\cos\phi = \frac{\langle\vec{a},\vec{b}\rangle}{|\vec{a}||\vec{b}|} = \frac{xt+yz}{\sqrt{x^2+y^2}\sqrt{t^2+z^2}}$$

The expression in 2-dimensional euclidean vector algebra for the magnitude of the cross product of two vectors is:

$$\sin\phi = \frac{|\vec{a}\times\vec{b}|}{|\vec{a}||\vec{b}|} = \frac{xz-yt}{\sqrt{x^2+y^2}\sqrt{t^2+z^2}}$$

Thus the inner product and the exterior product have fallen out of the algebra of the euclidean complex numbers as the way of calculating the angle between two vectors.

---

[22] The term exterior product is taken from the Clifford algebras and the Grassman algebras.

We refer to the operation we have just shown as the angle product of two complex numbers, and we denote it with the symbol, $\odot$.

Definition:

The Angle Product

The angle product of two matrices (within the same natural algebra) is the conjugate of one (either) matrix multiplied by the other matrix.

$$A \odot B = \overline{A}B$$

where the line atop the matrix denotes the conjugate of that matrix[23].

In the case of the euclidean complex numbers, we have:

$$\begin{bmatrix} x & y \\ -y & x \end{bmatrix} \odot \begin{bmatrix} t & z \\ -z & t \end{bmatrix} = \begin{bmatrix} x & -y \\ y & x \end{bmatrix}\begin{bmatrix} t & z \\ -z & t \end{bmatrix} = \begin{bmatrix} \begin{bmatrix} x \\ y \end{bmatrix} \cdot \begin{bmatrix} t \\ z \end{bmatrix} & \begin{bmatrix} x \\ y \end{bmatrix} \wedge \begin{bmatrix} t \\ z \end{bmatrix} \\ -\begin{bmatrix} x \\ y \end{bmatrix} \wedge \begin{bmatrix} t \\ z \end{bmatrix} & \begin{bmatrix} x \\ y \end{bmatrix} \cdot \begin{bmatrix} t \\ z \end{bmatrix} \end{bmatrix}$$

where we have used the wedge to indicate the exterior product. We note that:

$$\begin{bmatrix} a \\ b \end{bmatrix} \wedge \begin{bmatrix} c \\ d \end{bmatrix} = -\begin{bmatrix} c \\ d \end{bmatrix} \wedge \begin{bmatrix} a \\ b \end{bmatrix}$$

The angle product of two vectors is a vector.
We have:

$$[\phi] \odot [\theta] = \overline{[\theta] \odot [\phi]}$$
$$(\alpha[\phi] + \beta[\lambda]) \odot [\theta] = \alpha[\phi] \odot [\theta] + \beta[\lambda] \odot [\theta]$$
$$[\phi] \odot [\phi] = [\det[\phi]] > 0$$

Thus the angle product operation satisfies the axioms of an inner product[24], with the exterior product thrown in.

The reader should note that the exterior product cannot be the cross product. The space we have found within the euclidean complex numbers is 2-dimensional euclidean space; it has no third dimension.

The Clifford Algebra[25] View:

Clifford algebraists take the view (kept secret from undergraduates) that the cross product operation does not produce a vector but instead produces a bi-vector. A bi-vector is an area (a 2-dimensional plane of finite extent). However, in 3-dimensional space, to every 2-dimensional plane, there is a corresponding vector (perpendicular to the plane) up to direction (which is arbitrary). Thus, in 3-dimensional space, planes are dual to

---

[23] The conjugate of an algebraic matrix form is the adjoint of that matrix form.
[24] We thus have a complete Hilbert space – which, in the case of the euclidean complex numbers, is nothing new.
[25] In spite of the names, Clifford algebras do not satisfy the algebraic axioms and are thus not algebras.

vectors.[26] It is this duality that allows teachers to lead undergraduates into thinking that the output of the cross product operation is a vector. The particular vector depends upon the metric of the 3-dimensional space.

The Natural Algebra View:
In the natural algebras, the exterior product is neither a vector nor a bi-vector. The inner product and the exterior product are components of a vector.

Within the algebraic matrix form of the euclidean complex numbers, there are two parts. The real part and the 'imaginary' part respectively:

$$\begin{bmatrix} a & 0 \\ 0 & a \end{bmatrix}, \begin{bmatrix} 0 & b \\ -b & 0 \end{bmatrix}$$

Multiplication by the imaginary part rotates a vector, and it is exponentiation of the imaginary matrix that produces the rotation matrix. Thus, the exterior product is associated with rotation in the euclidean plane. We now see why, in physics, angular velocity is associated with the cross product[27].

What is 2-dimensional euclidean space?
2-dimensional euclidean space is often viewed as being two 1-dimensional spaces fixed together at right angles. The 2-dimensional euclidean space within the euclidean complex numbers is not two 1-dimensional spaces fixed together at right angles. If it were, then $\hat{i} \times \hat{i} = \hat{i}$ in the same way that $1 \times 1 = 1$. 2-dimensional euclidean space is a 2-dimensional space, which is a different thing from two 1-dimensional spaces fixed together.

The traditional view of (2-dimensional) vectors is that a vector is an ordered pair of real numbers. This is the view that 2-dimensional space is two 1-dimension spaces fixed together. The view given above is that a vector is a complex number. The complex numbers are such that the real part and the imaginary part can be multiplied together within a single algebra. This allows us to form the inner product and the exterior product legitimately within the algebra. The traditional view of vectors as a pair of real numbers does not allow the formation of the inner product and exterior product unless one allows the elements of two separate algebras to be multiplied together[28]. This is like breeding horses and cows together.

Within multilinear algebra is the concept of invariants. An invariant is something that remains the same under change of basis. The angle between two vectors (subtended at the origin) is such an invariant. The length of a vector is another invariant. The angle product is a way of calculating an invariant (the angle). The determinant of a matrix is a way of calculating another invariant (the length).

---

[26]This duality is known as the Hodge duality.
[27] When we do vector calculus, we get Div in place of the inner product and Curl in place of the exterior product. Of course, Curl is to do with circular motion.
[28] Which is done all the time in multi-linear algebra. Your author questions the legitimacy of this behavior.

Summary:
The euclidean complex numbers are the algebraic vector product algebra of 2-dimensional euclidean space. Within this algebra the algebraic vector product (matrix multiplication) creates the space. The metric and the inner product (and the wedge product) are simply calculative procedures within what is already a geometric space. This is very different from the conventional mathematical procedure of endowing a space with structure by putting a metric and an inner product on to it.
In due course, we will see that all the natural algebras are algebraic vector product algebras, and that they are the only such algebras.

An Aside:

The Clifford Algebra of 2-dimensional Euclidean Space

Multiplicative distributivity is a basic property of numbers. The inner product of two vectors is component-wise multiplication and is not distributive. If we multiply a vector by itself distributively, we get:

$$\left(a_1\vec{e}_1 + a_2\vec{e}_2\right)\left(a_1\vec{e}_1 + a_2\vec{e}_2\right) = a_1{}^2\vec{e}_1\vec{e}_1 + a_1a_2\vec{e}_1\vec{e}_2 + a_2a_1\vec{e}_2\vec{e}_1 + a_2{}^2\vec{e}_2\vec{e}_2$$

Setting this equal to the inner product, $a_1{}^2 + a_2{}^2$, (the cross product of a vector with itself is zero) compels us to set:

$$\left\{\vec{e}_1\vec{e}_1 = \vec{e}_2\vec{e}_2 = 1, \left|\vec{e}_1\vec{e}_2\right| = -\left|\vec{e}_2\vec{e}_1\right|\right\}$$

Using these relations and multiplying two different vectors together distributively gives the Cl_2[29] Clifford product:

$$\left(a_1\vec{e}_1 + a_2\vec{e}_2\right)\left(b_1\vec{e}_1 + b_2\vec{e}_2\right) = a_1b_1\vec{e}_1\vec{e}_1 + a_1b_2\vec{e}_1\vec{e}_2 + a_2b_1\vec{e}_2\vec{e}_1 + a_2b_2\vec{e}_2\vec{e}_2$$

$$= \left(a_1b_1 + a_2b_2\right) + \left(a_1b_2 - a_2b_1\right)\vec{e}_1\vec{e}_2$$

$$= \begin{bmatrix} a_1 \\ a_2 \end{bmatrix} \cdot \begin{bmatrix} b_1 \\ b_2 \end{bmatrix} + \begin{bmatrix} a_1 \\ a_2 \end{bmatrix} \wedge \begin{bmatrix} b_1 \\ b_2 \end{bmatrix}$$

Thus, the Clifford product produces both the inner product and the exterior product. The $\vec{e}_1\vec{e}_2$ object is a bi-vector. The Clifford product is not an algebraic multiplication because the product of two vectors is not a vector.

---

[29] Cl_2 is the name of the Clifford algebra of 2-dimensional euclidean space.

Section 2 : Chapter 5

Logarithms and Powers

Logarithms:

The inverse of the exponential function of 2×2 matrices we call the logarithm function of 2×2 matrices. Like the exponential, it is a function that takes 2×2 matrices for its arguments. We denote it by $\log m_{[e]}$. Because it is the inverse of the exponential, then, if and only if A & B commute, we have:

$$\log m_{[e]}(AB) = \log m_{[e]}(A) + \log m_{[e]}(B)$$

For the euclidean complex numbers, we have:

$$\begin{bmatrix} a & b \\ -b & a \end{bmatrix} = \begin{bmatrix} r & 0 \\ 0 & r \end{bmatrix}\begin{bmatrix} \cos\theta & \sin\theta \\ -\sin\theta & \cos\theta \end{bmatrix}$$

$$\log m_{[e]}\left(\begin{bmatrix} a & b \\ -b & a \end{bmatrix}\right) = \log m_{[e]}\left(\begin{bmatrix} r & 0 \\ 0 & r \end{bmatrix}\begin{bmatrix} \cos\theta & \sin\theta \\ -\sin\theta & \cos\theta \end{bmatrix}\right)$$

$$= \log m_{[e]}\left(\begin{bmatrix} r & 0 \\ 0 & r \end{bmatrix}\right) + \log m_{[e]}\left(\begin{bmatrix} \cos\theta & \sin\theta \\ -\sin\theta & \cos\theta \end{bmatrix}\right)$$

$$= \log m_{[e]}\left(\begin{bmatrix} r & 0 \\ 0 & r \end{bmatrix}\right) + \log m_{[e]}\left(\begin{bmatrix} \cos(\theta+2n\pi) & \sin(\theta+2n\pi) \\ -\sin(\theta+2n\pi) & \cos(\theta+2n\pi) \end{bmatrix}\right)$$

$$= \log m_{[e]}\left(\begin{bmatrix} r & 0 \\ 0 & r \end{bmatrix}\right) + \log m_{[e]}\left(\exp\left(\begin{bmatrix} 0 & (\theta+2n\pi) \\ -\theta+2n\pi & 0 \end{bmatrix}\right)\right)$$

$$= \log m_{[e]}\left(\begin{bmatrix} r & 0 \\ 0 & r \end{bmatrix}\right) + \begin{bmatrix} 0 & \theta+2n\pi \\ -(\theta+2n\pi) & 0 \end{bmatrix}$$

And so we have that the logarithm of a euclidean complex number is multi-valued.

For comparison, dropping from matrix notation into the usual notation:

$$\log m_{[e]}\left(\begin{bmatrix} r & 0 \\ 0 & r \end{bmatrix}\right) + \log m_{[e]}\left(\begin{bmatrix} \cos\theta & \sin\theta \\ -\sin\theta & \cos\theta \end{bmatrix}\right) \equiv \log_e r + \log_e\left(\cos\theta + \hat{i}\sin\theta\right)$$

$$\equiv \log_e r + \log_e\left(\frac{e^{i\theta}+e^{-i\theta}}{2} + \hat{i}\frac{e^{i\theta}-e^{-i\theta}}{2\hat{i}}\right)$$

$$\equiv \log_e r + \log_e\left(e^{i\theta}\right)$$

$$\equiv \log_e r + \log_e\left(e^{\hat{i}(\theta+2n\pi)}\right)$$

$$\equiv \log_e r + \hat{i}(\theta+2n\pi)$$

With which the reader might be more familiar.

Powers:

    We can use the logarithm to calculate exponentials.
For example, calculate:

$$\begin{bmatrix} 0 & b \\ -b & 0 \end{bmatrix}^{\begin{bmatrix} 0 & 1 \\ -1 & 0 \end{bmatrix}}$$

We set:

$$\begin{bmatrix} 0 & b \\ -b & 0 \end{bmatrix} = \exp\left(\log m_{[e]}\left(\begin{bmatrix} 0 & b \\ -b & 0 \end{bmatrix} \right) \right)$$

$$\begin{bmatrix} 0 & b \\ -b & 0 \end{bmatrix}^{\begin{bmatrix} 0 & 1 \\ -1 & 0 \end{bmatrix}} = \exp\left(\log m_{[e]}\left(\begin{bmatrix} 0 & b \\ -b & 0 \end{bmatrix} \right) \right)^{\begin{bmatrix} 0 & 1 \\ -1 & 0 \end{bmatrix}}$$

$$= \exp\left(\begin{bmatrix} 0 & 1 \\ -1 & 0 \end{bmatrix} \log m_{[e]}\left(\begin{bmatrix} 0 & b \\ -b & 0 \end{bmatrix} \right) \right)$$

The equations $\{0 = r\cos\theta, b = r\sin\theta\}$ have solutions

$$\left\{ \left\{ r = b, \theta = \frac{\pi}{2} \right\}, \left\{ r = -b, \theta = -\frac{\pi}{2} \right\} \right\}.$$

We take the positive solution:

$$\begin{bmatrix} 0 & b \\ -b & 0 \end{bmatrix}^{\begin{bmatrix} 0 & 1 \\ -1 & 0 \end{bmatrix}} = \exp\left(\begin{bmatrix} 0 & 1 \\ -1 & 0 \end{bmatrix} \left(\log m_{[e]}\left(\begin{bmatrix} b & 0 \\ 0 & b \end{bmatrix} \right) + \begin{bmatrix} 0 & \frac{\pi}{2} + 2n\pi \\ -\left(\frac{\pi}{2} + 2n\pi\right) & 0 \end{bmatrix} \right) \right)$$

$$= \exp\left(\begin{bmatrix} 0 & 1 \\ -1 & 0 \end{bmatrix} \log m_{[e]}\left(\begin{bmatrix} b & 0 \\ 0 & b \end{bmatrix} \right) \right) \exp\left(\begin{bmatrix} 0 & 1 \\ -1 & 0 \end{bmatrix} \begin{bmatrix} 0 & \frac{\pi}{2} + 2n\pi \\ -\left(\frac{\pi}{2} + 2n\pi\right) & 0 \end{bmatrix} \right)$$

$$= \exp\left(\begin{bmatrix} 0 & \log_e b \\ -\log_e b & 0 \end{bmatrix} \right) \exp\left(\begin{bmatrix} -\left(\frac{\pi}{2} + 2n\pi\right) & 0 \\ 0 & -\left(\frac{\pi}{2} + 2n\pi\right) \end{bmatrix} \right)$$

$$= \begin{bmatrix} \cos(\log_e b) & \sin(\log_e b) \\ -\sin(\log_e b) & \cos(\log_e b) \end{bmatrix} \exp\left(\begin{bmatrix} -\left(\frac{\pi}{2} + 2n\pi\right) & 0 \\ 0 & -\left(\frac{\pi}{2} + 2n\pi\right) \end{bmatrix} \right)$$

If $b = 1$, this is:

$$\begin{bmatrix} 0 & 1 \\ -1 & 0 \end{bmatrix}^{\begin{bmatrix} 0 & 1 \\ -1 & 0 \end{bmatrix}} = \begin{bmatrix} 1 & 0 \\ 0 & 1 \end{bmatrix} \exp\left(\begin{bmatrix} -\left(\dfrac{\pi}{2}+2n\pi\right) & 0 \\ 0 & -\left(\dfrac{\pi}{2}+2n\pi\right) \end{bmatrix}\right)$$

Which, written in the usual euclidean complex notation, is the familiar:

$$\hat{i}^{\hat{i}} = e^{-\frac{\pi}{2}+2n\pi}$$

We note that the exponential is multi-valued.

Section 2 : Chapter 6

Vector Calculus in 2-dimensional Euclidean Space

Conventional multilinear algebra recognizes two types of vectors, covariant vectors and contravariant vectors. Contravariant vectors are just normal vectors, but covariant vectors are linear functions that act on a vector. To add to the confusion, covariant vectors are also known as linear functions, linear functionals, co-vectors, vectors (sloppily), and 1-forms. Covariant vectors (1-forms) are written as row vectors:

$$[a \quad b]$$

Contravariant vectors (normal vectors) are written as column vectors:

$$\begin{bmatrix} x \\ y \end{bmatrix}$$

The 1-form (covariant vector) acts upon the contravariant vector (normal vector), by matrix multiplication, to produce a real number:

$$[a \quad b]\begin{bmatrix} x \\ y \end{bmatrix} = ax + by \in \mathbb{R}$$

This is the dot product of course (which is an invariant of the space).

Conventionally, the differential of a function, $f(a,b)$, is a 1-form (covariant vector):

$$df = \begin{bmatrix} \dfrac{\partial f}{\partial a}, & \dfrac{\partial f}{\partial b} \end{bmatrix}$$

This will act upon a contravariant vector to produce a real number:

$$\begin{bmatrix} \dfrac{\partial f}{\partial a} & \dfrac{\partial f}{\partial b} \end{bmatrix}\begin{bmatrix} v^1 \\ v^2 \end{bmatrix} = \dfrac{\partial f}{\partial a}v^1 + \dfrac{\partial f}{\partial b}v^2 \in \mathbb{R}$$

Aside:

Usual Notation

The basis 1-forms are, confusingly, written as $\{[dx \quad 0] \quad [0 \quad dy]\}$. The differential is often presented as a 1-form multiplied by the 1-form basis vectors and summed together as if the differential was a real number:

$$df = \dfrac{\partial f}{\partial a}da + \dfrac{\partial f}{\partial b}db$$

Doing vectors within an algebraic field considerable simplifies all the above. An essential part of an algebra is the product of two elements of that algebra where both elements are of the same nature. So it is that, when we do vectors within an algebraic field, there is no distinction between different types of vectors. In particular, there is no distinction between contravariant vectors and covariant vectors, and we can call them all just vectors – which is lovely.

A function upon a space is a scalar field upon that space. In matrix form, the function (scalar field) is:

$$\begin{bmatrix} f(a,b) & 0 \\ 0 & f(a,b) \end{bmatrix}$$

The differential is formed by taking the partial derivatives. The matrix configuration identifies these partial derivatives as components of a vector, and so we do not need to keep them separate:

$$d\begin{bmatrix} f(a,b) & 0 \\ 0 & f(a,b) \end{bmatrix} = \left[\begin{bmatrix} \dfrac{\partial f}{\partial a} & 0 \\ 0 & \dfrac{\partial f}{\partial a} \end{bmatrix}, \begin{bmatrix} 0 & -\dfrac{\partial f}{\partial b} \\ \dfrac{\partial f}{\partial b} & 0 \end{bmatrix} \right] = \begin{bmatrix} \dfrac{\partial f}{\partial a} & -\dfrac{\partial f}{\partial b} \\ \dfrac{\partial f}{\partial b} & \dfrac{\partial f}{\partial a} \end{bmatrix}$$

Of course, the differential is the gradient of the scalar field, $grad(f)$, which is a vector, which is what we have. A vector field is more than one function upon a space:

$$\begin{bmatrix} F_a(a,b) & F_b(a,b) \\ -F_b(a,b) & F_a(a,b) \end{bmatrix}$$

The differential of a vector field is:

$$d\begin{bmatrix} F_a(a,b) & F_b(a,b) \\ F_b(a,b) & F_a(a,b) \end{bmatrix} = \left[\begin{bmatrix} \dfrac{\partial F_a(a,b)}{\partial a} & \dfrac{\partial F_b(a,b)}{\partial a} \\ -\dfrac{\partial F_b(a,b)}{\partial a} & \dfrac{\partial F_a(a,b)}{\partial a} \end{bmatrix} \begin{bmatrix} \dfrac{\partial F_b(a,b)}{\partial b} & -\dfrac{\partial F_a(a,b)}{\partial b} \\ \dfrac{\partial F_a(a,b)}{\partial b} & \dfrac{\partial F_b(a,b)}{\partial b} \end{bmatrix} \right]$$

$$= \begin{bmatrix} \dfrac{\partial F_a(a,b)}{\partial a} + \dfrac{\partial F_b(a,b)}{\partial b} & \dfrac{\partial F_b(a,b)}{\partial a} - \dfrac{\partial F_a(a,b)}{\partial b} \\ \dfrac{\partial F_a(a,b)}{\partial b} - \dfrac{\partial F_b(a,b)}{\partial a} & \dfrac{\partial F_a(a,b)}{\partial a} + \dfrac{\partial F_b(a,b)}{\partial b} \end{bmatrix}$$

$$= \begin{bmatrix} div(F) & curl(F) \\ -curl(F) & div(F) \end{bmatrix}$$

Differentiating a scalar field twice with respect to both variables gives:

$$\begin{bmatrix} \dfrac{\partial^2 f}{\partial a^2} + \dfrac{\partial^2 f}{\partial b^2} & 0 \\ 0 & \dfrac{\partial^2 f}{\partial a^2} + \dfrac{\partial^2 f}{\partial b^2} \end{bmatrix}$$

which is the Laplacian for a scalar field.

The second differential of a vector field is:

$$\begin{bmatrix} \dfrac{\partial^2 F_a}{\partial a^2} + \dfrac{\partial^2 F_a}{\partial b^2} & \dfrac{\partial^2 F_b}{\partial a^2} + \dfrac{\partial^2 F_b}{\partial b^2} \\[4mm] -\left(\dfrac{\partial^2 F_b}{\partial a^2} + \dfrac{\partial^2 F_b}{\partial b^2} \right) & \dfrac{\partial^2 F_a}{\partial a^2} + \dfrac{\partial^2 F_a}{\partial b^2} \end{bmatrix}$$

which is the Laplacian for a vector field.

Introduction to Section 3

In this section, we seek to further familiarise the reader with doing algebra in matrix notation and to introduce the reader to another natural algebra – another form of complex numbers. We seek to prepare the reader for what comes after – an infinite number of new types of complex numbers each containing a new type of space.

Section 3 : Chapter 1

The Hyperbolic Complex Numbers – Basics

(Study Numbers) - $\mathbb{S}$

Readers will be familiar with the euclidean complex numbers but might be wholly unfamiliar with the hyperbolic complex numbers. Hyperbolic complex numbers are also called Study numbers[30]. It might be, and your author is far from certain about this, that Study numbers are named after Eduard Study, a German mathematician working around the year 1900[31]. Your author has scanned through Eduard Study's works but has found nothing at all concerning these hyperbolic complex numbers,[32] and your author wonders if he is erroneously attributing them to Eduard Study. The fancy 'S' denoting them is an invention of your author. Having placed the credit where credit might be due, we will henceforward refer to this algebra as the hyperbolic complex numbers.

The hyperbolic complex numbers are an algebraic field. Thus, their algebraic status, with the restricted values, is equal to that of the euclidean complex numbers.[33]

The matrix representation of a hyperbolic complex number, $\left(a + \hat{r}b \right) : \hat{r} = \sqrt{+1}$, is:

$$\begin{bmatrix} a & b \\ b & a \end{bmatrix} : |a| > |b|$$

in which, as always within this work, the matrix elements are real numbers. The reader might be inclined to think $\hat{r} = -1$. However, the matrices:

$$\left\{ \begin{bmatrix} 0 & 1 \\ 1 & 0 \end{bmatrix}, \begin{bmatrix} -1 & 0 \\ 0 & -1 \end{bmatrix} \right\}$$

are not the same. The restrictions are imposed to guarantee non-singularity of the matrix[34].

---

[30] Your author discovered the name Study numbers in the book 'Clifford Algebras and Spinors' (page 24) by Pertti Lounesto - ISBN 0-521-00551-5. This is the only instance I know of them being mentioned anywhere, and only four sentences. Unfortunately, Pertti Lounesto has died.

[31] Readers seeking his works are directed to the web site *copac.ac.uk*.

[32] This might be because, although the mathematics is written in mathematics, the English is written in German in which your author is not fluent.

[33] The hyperbolic complex numbers are the even sub-algebra, $Cl_{1,1}^{+}$, of the Clifford algebra of $\mathbb{R}^{1,1}$ which is

$Cl_{1,1}$.

[34] We will shortly see that the space within this algebra is the space-time of special relativity. From that, it folows that we have just calculated the limiting velocity of the universe (speed of light?).

Addition is matrix addition, and multiplication is matrix multiplication. The conjugate is the adjoint matrix:

$$\begin{bmatrix} a & -b \\ -b & a \end{bmatrix}.$$

In non-matrix terms, hyperbolic complex numbers are similar to euclidean complex numbers, but, instead of having $\hat{i} = \sqrt{-1}$, they have $\hat{r} = \sqrt{+1}$. They have the non-matrix form $x + \hat{r}y : |x| > |y|$. Addition and multiplication are parallels of the euclidean complex numbers:

$$\left(a + \hat{r}b\right) + \left(x + \hat{r}y\right) = \left(a + x\right) + \hat{r}\left(b + y\right)$$

$$\left(a + \hat{r}b\right)\left(x + \hat{r}y\right) = \left(ax + \hat{r}^2 by\right) + \hat{r}\left(ay + bx\right) = \left(ax + by\right) + \hat{r}\left(ay + bx\right)$$

The conjugate of the hyperbolic complex number $x + \hat{r}y$ is $x - \hat{r}y$, and division is done with the conjugate as with euclidean complex numbers.

$$\frac{a + \hat{r}b}{x + \hat{r}y} = \frac{\left(a + \hat{r}b\right)\left(x - \hat{r}y\right)}{\left(x + \hat{r}y\right)\left(x - \hat{r}y\right)} = \frac{\left(a + \hat{r}b\right)\left(x - \hat{r}y\right)}{x^2 - y^2}$$

Such division is not defined when $|x| = |y|$, but these values are excluded values.

Hyperbolic complex numbers sit on a plane in a way similar to the way euclidean complex numbers sit on the Argand diagram, but the hyperbolic complex number plane is hyperbolic space instead of euclidean space. As visualized by our euclidean eyes, the hyperbolic complex number plane is split into two halves by the 45° lines, where $|x| = |y|$, and the hyperbolic complex numbers occupy two horizontally opposite quadrants. The real numbers lie on the horizontal x-axis.

Below is an Argand diagram for hyperbolic complex numbers. The hyperbolas are the 'real' co-ordinates.

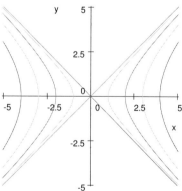

Argand diagram for the hyperbolic complex numbers

Section 3 : Addendum to Chapter 1

Proof that the two Notations of Hyperbolic Complex Numbers are Algebraically Isomorphic

---

Theorem

The algebra of matrices of the form $\begin{bmatrix} a & b \\ b & a \end{bmatrix} : |a| > |b|$ is isomorphic to the algebra of the

hyperbolic complex numbers, $a + \hat{r}b : |a| > |b|$.

Proof:

$$\phi : (\mathbb{S}, +, \times) \mapsto \left(M_{(2,\mathbb{R})}, +, \bullet\right) \quad : \quad \phi\left(a + \hat{r}b\right) \mapsto \begin{bmatrix} a & b \\ b & a \end{bmatrix}$$

i) $\quad \phi\left(\left(a + \hat{r}b\right) + \left(e + \hat{r}f\right)\right) \mapsto \begin{bmatrix} a+e & b+f \\ b+f & a+e \end{bmatrix} = \begin{bmatrix} a & b \\ b & a \end{bmatrix} + \begin{bmatrix} e & f \\ f & e \end{bmatrix}$

ii) $\quad \phi\left(\left(a + \hat{r}b\right) \times \left(e + \hat{r}f\right)\right) \mapsto \begin{bmatrix} ae+bf & af+be \\ af+be & ae+bf \end{bmatrix} = \begin{bmatrix} a & b \\ b & a \end{bmatrix} \bullet \begin{bmatrix} e & f \\ f & e \end{bmatrix}$

iii) $\quad \phi\left(\alpha\left(a + \hat{r}b\right)\right) \mapsto \begin{bmatrix} \alpha a & \alpha b \\ b & \alpha a \end{bmatrix} = \alpha \begin{bmatrix} a & b \\ b & a \end{bmatrix}$

Section 3 : Chapter 2

The Cauchy-Riemann Equations for Hyperbolic Complex Numbers

We consider functions of hyperbolic complex numbers. These are of the form:

$$f\left(\begin{bmatrix} a & b \\ b & a \end{bmatrix}\right) \rightarrow \begin{bmatrix} u(a,b) & v(a,b) \\ v(a,b) & u(a,b) \end{bmatrix}$$

We have:

$$\frac{\partial\left(\begin{bmatrix} u(a,b) & v(a,b) \\ v(a,b) & u(a,b) \end{bmatrix}\right)}{\partial\begin{bmatrix} a & 0 \\ 0 & a \end{bmatrix}} = \begin{bmatrix} 1 & 0 \\ 0 & 1 \end{bmatrix}\begin{bmatrix} \dfrac{\partial u}{\partial a} & 0 \\ 0 & \dfrac{\partial u}{\partial a} \end{bmatrix} + \begin{bmatrix} 0 & 1 \\ 1 & 0 \end{bmatrix}\begin{bmatrix} \dfrac{\partial v}{\partial a} & 0 \\ 0 & \dfrac{\partial v}{\partial a} \end{bmatrix} = \begin{bmatrix} \dfrac{\partial u}{\partial a} & \dfrac{\partial v}{\partial a} \\ \dfrac{\partial v}{\partial a} & \dfrac{\partial u}{\partial a} \end{bmatrix}$$

$$\frac{\partial\left(\begin{bmatrix} u(a,b) & v(a,b) \\ v(a,b) & u(a,b) \end{bmatrix}\right)}{\partial\begin{bmatrix} 0 & b \\ b & 0 \end{bmatrix}} = \begin{bmatrix} 1 & 0 \\ 0 & 1 \end{bmatrix}\begin{bmatrix} \dfrac{\partial u}{\partial b} & 0 \\ 0 & \dfrac{\partial u}{\partial b} \end{bmatrix} + \begin{bmatrix} 0 & 1 \\ 1 & 0 \end{bmatrix}\begin{bmatrix} \dfrac{\partial v}{\partial b} & 0 \\ 0 & \dfrac{\partial v}{\partial b} \end{bmatrix} = \begin{bmatrix} \dfrac{\partial v}{\partial b} & \dfrac{\partial u}{\partial b} \\ \dfrac{\partial u}{\partial b} & \dfrac{\partial v}{\partial b} \end{bmatrix}$$

Note: Because of the restrictions on the values of a & b, $\begin{bmatrix} 0 & 1 \\ 1 & 0 \end{bmatrix}$ is not a hyperbolic complex number. However, above, it is used as part of an expression that is a hyperbolic complex number.

The Cauchy Riemann equations of this algebra are:

$$\frac{\partial u}{\partial a} = \frac{\partial v}{\partial b} \quad \& \quad \frac{\partial u}{\partial b} = \frac{\partial v}{\partial a}$$

Section 3 : Chapter 3

The Polar Form of Hyperbolic Complex Numbers

Consider:

$$\exp\left(\begin{bmatrix} a & b \\ b & a \end{bmatrix}\right) = \begin{bmatrix} e^a & 0 \\ 0 & e^a \end{bmatrix} \begin{bmatrix} 1 + \dfrac{b^2}{2!} + \dfrac{b^4}{4!} + \dots & b + \dfrac{b^3}{3!} + \dfrac{b^5}{5!} + \dots \\ b + \dfrac{b^3}{3!} + \dfrac{b^5}{5!} + \dots & 1 + \dfrac{b^2}{2} + \dfrac{b^4}{4!} + \dots \end{bmatrix}$$

$$= \begin{bmatrix} e^a & 0 \\ 0 & e^a \end{bmatrix} \begin{bmatrix} \cosh b & \sinh b \\ \sinh b & \cosh b \end{bmatrix}$$

$$= \begin{bmatrix} r & 0 \\ 0 & r \end{bmatrix} \begin{bmatrix} \cosh b & \sinh b \\ \sinh b & \cosh b \end{bmatrix}$$

We have a length matrix and the rotation matrix of 2-dimensional hyperbolic space. The trace of the b-sub-matrix is zero, and so the determinant of the rotation matrix is unity, and we have the identity: $\cosh^2 b - \sinh^2 b = 1$.

We have:

$$\begin{bmatrix} x & y \\ y & x \end{bmatrix} = \begin{bmatrix} r & 0 \\ 0 & r \end{bmatrix} \begin{bmatrix} \cosh b & \sinh b \\ \sinh b & \cosh b \end{bmatrix}$$

$$\det\left(\begin{bmatrix} x & y \\ y & x \end{bmatrix}\right) = \det\left(\begin{bmatrix} r & 0 \\ 0 & r \end{bmatrix} \begin{bmatrix} \cosh b & \sinh b \\ \sinh b & \cosh b \end{bmatrix}\right)$$

$$x^2 - y^2 = r^2$$

Leading to the distance function[35] of 2-dimensional hyperbolic space:

$$r = \sqrt{x^2 - y^2}$$

We note that the polar form can never have determinant equal to zero except for the zero matrix. The polar form does not need to be restricted to guarantee non-singularity. The reader should note that $|\cosh \chi| > |\sinh \chi| \quad \forall \chi$.

Important Point:
We do not need to restrict the algebra of hyperbolic complex numbers in polar form. This means that this algebra consists of the unrestricted matrices that are equal to polar form matrices. That is, for unrestricted values of $\{a,b\}$, the algebra is the set of matrices of the form:

$$\begin{bmatrix} x & y \\ y & x \end{bmatrix} \text{ where } \{x = e^a \cosh b, \, y = e^a \sinh b\}$$

---

[35] Technically, this function is not a metric – it violates the triangle inequality, axiom 3. See Appendix 1.

40

We make the same points that we made when we found the polar form of the euclidean complex numbers in Section 2.

1) The trigonometric functions associated with 2-dimensional hyperbolic space fall out of the algebra of the hyperbolic complex numbers.

2) The rotation matrix of 2-dimensional hyperbolic space falls out of the algebra of the hyperbolic complex numbers.

3) The distance function of 2-dimensional hyperbolic space falls out of the algebra of the hyperbolic complex numbers.

We take the view that this is sufficient to say that 2-dimensional hyperbolic space is contained within the algebra of the hyperbolic complex numbers.

Theorem:
2-dimensional hyperbolic space is contained within the algebra of the hyperbolic complex numbers.
Proof:
See Above

With this theorem, we have effectively unified the geometry of 2-dimensional hyperbolic space and the algebra of the hyperbolic complex numbers.

2-dimensional hyperbolic space (otherwise known as Minkowski space-time) is the space-time of special relativity. Special relativity is a 2-dimensional theory, and the commonly given space-time distance function, $\tau^2 = t^2 - x^2 - y^2 - z^2$, is incorrect. The correct form is: $\tau^2 = t^2 - x^2$. The extra terms are added in a 'fudge' designed to take account of the three spatial dimensions we perceive. This algebra is an algebraic field because we restricted the cartesian values of the variables as: $\frac{b}{a} < 1$. Since this is space-time, $\frac{b}{a} = v$, and we have a limiting velocity. Unfortunately, we have no concept of electricity or magnetism within this algebra, so we can not identify this limiting velocity with the velocity of light. However, starting from no more than the existence of the real numbers, we have calculated a physical constant. It comes out as 1 because there are no man-made units in the real numbers.

The hyperbolic complex numbers have polar form based on the hyperbolic trigonometric functions *cosh* & *sinh*; these are not periodic functions. The euclidean complex numbers have polar form based on the euclidean trigonometric functions *cos* & *sin*; these are periodic functions. It is the periodic nature of the euclidean trigonometric functions that 'causes' the multi-valued nature of many euclidean complex functions. The non-periodic nature of the hyperbolic trigonometric functions leads to the non-periodic nature of functions within this algebra.

Section 3 : Chapter 4

Logarithms and Powers
(and the Euler Relations for Sinh & Cosh)

The Hyperbolic Euler Relations:

The Euler relations for the hyperbolic trigonometric functions are usually given as:

$$\cosh \chi = \frac{e^{\chi} + e^{-\chi}}{2}, \quad \sinh \chi = \frac{e^{\chi} - e^{-\chi}}{2}$$

We consider:

$$e^{\hat{r}\chi} = 1 + \hat{r}\chi + \frac{\hat{r}^{2}\chi^{2}}{2!} + \frac{\hat{r}^{3}\chi^{3}}{3!} + \dots$$

where $\hat{r} = \sqrt{+1} \neq \pm 1$. This leads to:

$$e^{\hat{r}\chi} = \cosh \chi + \hat{r}\sinh \chi$$

Similarly:

$$e^{-\hat{r}\chi} = 1 - \hat{r}\chi + \frac{\hat{r}^{2}\chi^{2}}{2!} - \frac{\hat{r}^{3}\chi^{3}}{3!} + \dots$$

This leads to:

$$e^{-\hat{r}\chi} = \cosh \chi - \hat{r}\sinh \chi$$

Combining these results gives:

$$\cosh \chi = \frac{e^{\hat{r}\chi} + e^{-\hat{r}\chi}}{2}$$

$$\sinh \chi = \frac{e^{\hat{r}\chi} - e^{-\hat{r}\chi}}{2\hat{r}}$$

The reader might like to compare these to the euclidean Euler relations:

$$\cos \theta = \frac{e^{\hat{i}\theta} + e^{-\hat{i}\theta}}{2}, \quad \sin \theta = \frac{e^{\hat{i}\theta} - e^{-\hat{i}\theta}}{2\hat{i}}$$

No more need be said.

For the hyperbolic complex numbers, we have:

$$\begin{bmatrix} a & b \\ b & a \end{bmatrix} = \begin{bmatrix} h & 0 \\ 0 & h \end{bmatrix} \begin{bmatrix} \cosh \chi & \sinh \chi \\ \sinh \chi & \cosh \chi \end{bmatrix}$$

$$\log m_{[e]}\left(\begin{bmatrix} a & b \\ b & a \end{bmatrix} \right) = \log m_{[e]}\left(\begin{bmatrix} h & 0 \\ 0 & h \end{bmatrix} \begin{bmatrix} \cosh \chi & \sinh \chi \\ \sinh \chi & \cosh \chi \end{bmatrix} \right)$$

$$= \log m_{[e]}\left(\begin{bmatrix} h & 0 \\ 0 & h \end{bmatrix} \right) + \log m_{[e]}\left(\begin{bmatrix} \cosh \chi & \sinh \chi \\ \sinh \chi & \cosh \chi \end{bmatrix} \right)$$

$$= \log m_{[e]}\left(\begin{bmatrix} h & 0 \\ 0 & h \end{bmatrix} \right) + \log m_{[e]}\left(\exp\left(\begin{bmatrix} 0 & \chi \\ \chi & 0 \end{bmatrix} \right) \right)$$

$$= \log m_{[e]}\left(\begin{bmatrix} h & 0 \\ 0 & h \end{bmatrix} \right) + \begin{bmatrix} 0 & \chi \\ \chi & 0 \end{bmatrix}$$

And, so, we have that the logarithm of a hyperbolic complex number is not multi-valued.

Theorem:
Logarithms of hyperbolic complex numbers are not multi-valued.
Proof:
See Above

Dropping from matrix notation into non-matrix notation:

$$\log m_{[e]}\left(\begin{bmatrix} h & 0 \\ 0 & h \end{bmatrix} \right) + \log m_{[e]}\left(\begin{bmatrix} \cosh \chi & \sinh \chi \\ \sinh \chi & \cosh \chi \end{bmatrix} \right) \equiv \log_e h + \log_e \left(\cosh \chi + \hat{r} \sinh \chi \right)$$

$$\equiv \log_e h + \log_e \left(\frac{e^{r\chi} + e^{-r\chi}}{2} + \hat{r}\frac{e^{r\chi} - e^{-r\chi}}{2\hat{r}} \right)$$

$$\equiv \log_e h + \log_e \left(e^{r\chi} \right)$$

$$\equiv \log_e h + \hat{r}\chi$$

Section 3 : Chapter 5

Odd Parity Matrices, Quaternions, and Dihedralions

The euclidean complex numbers are a sub-algebra of the quaternions. We consider the hyperbolic complex numbers in this light.

Odd Parity Matrices:

Of all the 2×2 matrices with characterizations $\{a_{11} \in \mathbb{R}, a_{21} \in \mathbb{R}, a_{21} = \pm a_{12}, a_{22} = \pm a_{11}\}$, there is only one matrix with zero minus signs. That one is:

1) $\mathbb{S} \simeq \begin{bmatrix} a & b \\ b & a \end{bmatrix} \triangleq \begin{bmatrix} -a & -b \\ -b & -a \end{bmatrix} \triangleq \begin{bmatrix} -a & b \\ b & -a \end{bmatrix} \triangleq \begin{bmatrix} a & -b \\ -b & a \end{bmatrix}$

where we have used the $\triangleq$ sign to indicate equality of matrix form. We have found that the algebra of these matrices, when restricted to $|a| > |b|$, is the algebra of the hyperbolic complex numbers. There are two 2×2 such matrices with one minus quantity. Those two are:

2) $\mathbb{C} \simeq \begin{bmatrix} a & -b \\ b & a \end{bmatrix} \triangleq \begin{bmatrix} a & b \\ -b & a \end{bmatrix} \triangleq \begin{bmatrix} -a & -b \\ b & -a \end{bmatrix} \triangleq \begin{bmatrix} -a & b \\ -b & -a \end{bmatrix}$

3) $\mathbb{OP}_\mathbb{C} = \begin{bmatrix} a & b \\ b & -a \end{bmatrix} \triangleq \begin{bmatrix} -a & b \\ b & a \end{bmatrix} \triangleq \begin{bmatrix} a & -b \\ -b & -a \end{bmatrix} \triangleq \begin{bmatrix} -a & -b \\ -b & a \end{bmatrix}$

The first of these is the euclidean complex numbers. To the second, we have given the appellation $\mathbb{OP}_\mathbb{C}$[36]. There is only one matrix with two minus quantities. That one is:

4) $\mathbb{OP}_\mathbb{S} = \begin{bmatrix} a & -b \\ b & -a \end{bmatrix} \triangleq \begin{bmatrix} -a & b \\ -b & a \end{bmatrix} \triangleq \begin{bmatrix} -a & -b \\ b & a \end{bmatrix} \triangleq \begin{bmatrix} a & b \\ -b & -a \end{bmatrix}$

To this matrix form we have given the appellation $\mathbb{OP}_\mathbb{S}$. The matrices with three or four minus quantities are included in the above.

Let us first consider the third of these matrix types:

$$\mathbb{OP}_\mathbb{C} = \begin{bmatrix} a & b \\ b & -a \end{bmatrix}$$

When two of these matrices are multiplied together they form a euclidean complex number. When three of these matrices are multiplied together they return to their own form. When four of these matrices are multiplied together they form a euclidean complex number; and so it goes on.

$$\begin{bmatrix} a & b \\ b & -a \end{bmatrix}\begin{bmatrix} e & f \\ f & -e \end{bmatrix} = \begin{bmatrix} ae+bf & af-be \\ -(af-be) & ae+bf \end{bmatrix}$$

---

[36] Odd Parity

44

$$\begin{bmatrix} ae+bf & af-be \\ -(af-be) & ae+bf \end{bmatrix}\begin{bmatrix} m & n \\ n & -m \end{bmatrix}$$

$$=\begin{bmatrix} m(ae+bf)+n(af-be) & n(ae+bf)-m(af-be) \\ n(ae+bf)-m(af-be) & -(m(ae+bf)+n(af-be)) \end{bmatrix}$$

These matrices are not multiplicatively commutative, but, if the multiplying order is reversed, they multiply to the conjugate of the euclidean complex number.

$$\begin{bmatrix} e & f \\ f & -e \end{bmatrix}\begin{bmatrix} a & b \\ b & -a \end{bmatrix}=\begin{bmatrix} ae+bf & -(af-be) \\ af-be & ae+bf \end{bmatrix}$$

They square to a real number.

$$\begin{bmatrix} a & b \\ b & -a \end{bmatrix}\begin{bmatrix} a & b \\ b & -a \end{bmatrix}=\begin{bmatrix} a^2+b^2 & 0 \\ 0 & a^2+b^2 \end{bmatrix}$$

What kind of a creature is it whose product is a euclidean complex number and whose square is a real number? It is comprised of two 'bits', the a 'bit' and the b 'bit'. Thus, it is of the form: $a\hat{j}+b\hat{k}$. If we square such a creature, we get a positive real number, a^2+b^2:

$$a^2\hat{j}^2+b^2\hat{k}^2+ab\left(\hat{j}\hat{k}+\hat{k}\hat{j}\right)=a^2+b^2$$

Therefore, we conclude that $\hat{j}\hat{k}=-\hat{k}\hat{j}$ and that $\hat{j}^2=+1$ and that $\hat{k}^2=+1$.
Multiplied together, two different such creatures are:

$$\left(a\hat{j}+b\hat{k}\right)\left(e\hat{j}+f\hat{k}\right)=ae\hat{j}^2+bf\hat{k}^2+af\hat{j}\hat{k}+be\hat{k}\hat{j}$$

This, we know, is a euclidean complex number. The imaginary part is:

$$af\hat{j}\hat{k}+be\hat{k}\hat{j}=iy$$

Using the above, we conclude that $\hat{j}\hat{k}=\pm i$ and $\hat{k}\hat{j}=\mp i$.

Some Clifford Algebra

The Clifford algebra Cl_2 is a 4-dimensional algebra with basis: $\left\{1,\overline{e_1},\overline{e_2},\overline{e_1e_2}\right\}$.
These basis elements are algebraically isomorphic to the matrices:

$$\left\{1\cong\begin{bmatrix} 1 & 0 \\ 0 & 1 \end{bmatrix},\overline{e_1}\cong\begin{bmatrix} 1 & 0 \\ 0 & -1 \end{bmatrix},\overline{e_2}\cong\begin{bmatrix} 0 & 1 \\ 1 & 0 \end{bmatrix},\overline{e_1e_2}\cong\begin{bmatrix} 0 & 1 \\ -1 & 0 \end{bmatrix}\right\}$$

We have:

$$\mathbb{OP}_\mathbb{C}=\begin{bmatrix} a & b \\ b & -a \end{bmatrix}\cong a\overline{e_1}+b\overline{e_2}$$

$$\mathbb{C}=\begin{bmatrix} a & b \\ -b & a \end{bmatrix}\cong a+b\overline{e_1}\overline{e_2}$$

Let us now consider the algebra of the fourth of these matrix types:

$$\mathbb{OP}_S = \begin{bmatrix} a & -b \\ b & -a \end{bmatrix}$$

We shall consider only the matrices of this form that have $\left|a^2\right| > \left|b^2\right|$. When two of these matrices are multiplied together they form a hyperbolic complex number. When three of these matrices are multiplied together they return to their own form; and so it goes on.

$$\begin{bmatrix} a & -b \\ b & -a \end{bmatrix}\begin{bmatrix} e & -f \\ f & -e \end{bmatrix} = \begin{bmatrix} ae-bf & -(af-be) \\ -(af-be) & ae-bf \end{bmatrix}$$

$$\begin{bmatrix} ae-bf & -(af-be) \\ -(af-be) & ae-bf \end{bmatrix}\begin{bmatrix} m & -n \\ n & -m \end{bmatrix} =$$

$$\begin{bmatrix} m(ae-bf)-n(af-be) & -(n(ae-bf)-m(af-be)) \\ n(ae-bf)-m(af-be) & -(m(ae-bf)-n(af-be)) \end{bmatrix}$$

These matrices are not multiplicatively commutative, but, if the multiplying order is reversed, they multiply to the conjugate of the hyperbolic complex number.

$$\begin{bmatrix} e & -f \\ f & -e \end{bmatrix}\begin{bmatrix} a & -b \\ b & -a \end{bmatrix} = \begin{bmatrix} ae+bf & af-be \\ af-be & ae+bf \end{bmatrix}$$

They square to a positive real number.

$$\begin{bmatrix} a & -b \\ b & -a \end{bmatrix}\begin{bmatrix} a & -b \\ b & -a \end{bmatrix} = \begin{bmatrix} a^2-b^2 & 0 \\ 0 & a^2-b^2 \end{bmatrix}$$

We have:

$$a^2\hat{s}^2 + b^2\hat{t}^2 + ab\left(\hat{s}\hat{t}+\hat{t}\hat{s}\right) = a^2 - b^2$$

Therefore, we conclude that $\hat{s}\hat{t} = -\hat{t}\hat{s}$ and that $\hat{s}^2 = +1$ and that $\hat{t}^2 = -1$.
We have:

$$\left(a\hat{s}+b\hat{t}\right)\left(e\hat{s}+f\hat{t}\right) = ae\hat{s}^2 + bf\hat{t}^2 + af\hat{s}\hat{t} + be\hat{t}\hat{s}$$

This, we know, is a hyperbolic complex number. The 'imaginary' part is:

$$af\hat{s}\hat{t} + be\hat{t}\hat{s} = \hat{r}y$$

where we take $\hat{r} = \sqrt{+1}$. We conclude that $\hat{s}\hat{t} = \pm\hat{r}$ and $\hat{t}\hat{s} = \mp\hat{r}$.

More Clifford Algebra

The Clifford algebra $Cl_{1,1}$ is a 4-dimensional algebra with basis: $\left\{1,\overline{e}_1,\overline{e}_2,\overline{e_1e_2}\right\}$.

These basis elements are algebraically isomorphic to the matrices:

$$\left\{1 \cong \begin{bmatrix} 1 & 0 \\ 0 & 1 \end{bmatrix}, \overline{e}_1 \cong \begin{bmatrix} 0 & -1 \\ 1 & 0 \end{bmatrix}, \overline{e}_2 \cong \begin{bmatrix} 1 & 0 \\ 0 & -1 \end{bmatrix}, \overline{e_1e_2} \cong \begin{bmatrix} 0 & 1 \\ 1 & 0 \end{bmatrix}\right\}$$

We have:

$$OP_s = \begin{bmatrix} a & -b \\ b & -a \end{bmatrix} \cong a\overline{e_1} + b\overline{e_2}$$

$$S = \begin{bmatrix} a & b \\ b & a \end{bmatrix} \cong a + b\overline{e_1}\overline{e_2}$$

We will soon combine these intriguing odd parity matrices with complex numbers and hyperbolic complex numbers to make quaternions and dihedralions, but, before that, a small digression.

A Brief (and Conventional) Introduction to Quaternions - $\mathbb{H}$

The quaternions were discovered by the Irish mathematician William Hamilton. They were seen as an extension of complex numbers from two dimensions into four dimensions.

Quaternions are of the form: $q = a + \hat{i}b + \hat{j}c + \hat{k}d$. They are multiplicatively non-commutative, but they have all the other attributes of an algebraic field. Thus, they form a division algebra. We have $\hat{i}^2 = \hat{j}^2 = \hat{k}^2 = -1$ and the multiplicative relations $\hat{i}\hat{j} = \hat{k}, \hat{j}\hat{i} = -\hat{k}, \hat{j}\hat{k} = \hat{i}, \hat{k}\hat{j} = -\hat{i}, \hat{k}\hat{i} = \hat{j}, \hat{i}\hat{k} = -\hat{j}$. Addition and multiplication are what you would expect (taking account of the above multiplicative relations):

Addition:

$$\left(a + \hat{i}b + \hat{j}c + \hat{k}d\right) + \left(e + \hat{i}f + \hat{j}g + \hat{k}h\right) = (a+e) + \hat{i}(b+f) + \hat{j}(c+g) + \hat{k}(d+h)$$

Multiplication:

$$\left(a + \hat{i}b + \hat{j}c + \hat{k}d\right)\left(e + \hat{i}f + \hat{j}g + \hat{k}h\right) = a\left(e + \hat{i}f + \hat{j}g + \hat{k}h\right) + \hat{i}b\left(e + \hat{i}f + \hat{j}g + \hat{k}h\right) +$$
$$\hat{j}c\left(e + \hat{i}f + \hat{j}g + \hat{k}h\right) + \hat{k}d\left(e + \hat{i}f + \hat{j}g + \hat{k}h\right)$$
$$= ae + \hat{i}af + \hat{j}ag + \hat{k}ah + \hat{i}be + \hat{i}^2 bf + \hat{i}\hat{j}bg + \hat{i}\hat{k}bh +$$
$$\hat{j}ce + \hat{j}\hat{i}cf + \hat{j}^2 cg + \hat{j}\hat{k}ch + \hat{k}de + \hat{k}\hat{i}df + \hat{k}\hat{j}dg + \hat{k}^2 dh$$
$$= ae - bf - cg - dh + \hat{i}(af + be + ch - dg) +$$
$$\hat{j}(ag - bh + ce + df) + \hat{k}(ah + bg - cf + de)$$

The (conventional) conjugate of a quaternion $q = a + \hat{i}b + \hat{j}c + \hat{k}d$ is $q^* = a - \hat{i}b - \hat{j}c - \hat{k}d$, and division is done with the conjugate as with complex numbers.

$$\frac{\left(a + \hat{i}b + \hat{j}c + \hat{k}d\right)}{\left(e + \hat{i}f + \hat{j}g + \hat{k}h\right)} = \frac{\left(a + \hat{i}b + \hat{j}c + \hat{k}d\right)\left(e - \hat{i}f - \hat{j}g - \hat{k}h\right)}{\left(e + \hat{i}f + \hat{j}g + \hat{k}h\right)\left(e - \hat{i}f - \hat{j}g - \hat{k}h\right)}$$

$$= \frac{\left(a+\hat{i}b+\hat{j}c+\hat{k}d\right)\left(e-\hat{i}f-\hat{j}h-\hat{k}h\right)}{e^2+f^2+g^2+h^2}$$

The (conventional) norm of quaternions is $Norm = a^2+b^2+c^2+d^2$.

The quaternions are only a quarter of the story. We now relate the story in full. The odd parity matrices occur as 2×2 blocks within quaternion-like matrices. There are four types of quaternion-like algebras. We have:

Matrix Form 1:

$$\begin{bmatrix} \mathbb{C} & \mathbb{OP}_\mathbb{C} \\ \mathbb{OP}_\mathbb{C} & \mathbb{C} \end{bmatrix} \triangleq \begin{bmatrix} a & b & c & d \\ -b & a & d & -c \\ c & d & a & b \\ d & -c & -b & a \end{bmatrix}$$

This is the algebra of a euclidean complex number and an odd parity number set in the matrix form of a hyperbolic complex number. This matrix form is multiplicatively closed but not multiplicatively commutative. It is inversely closed with: $\det = \left(a^2+b^2-c^2-d^2\right)^2$ which is positive definite for all values of $\{a,b,c,d\}$. We take the norm to be the fourth root of the determinant.

In non-matrix notation, the number is: $a+\hat{i}b+\hat{j}c+\hat{k}d$. The conjugate, taken from the adjoint matrix, is:

$$a\left(a^2+b^2-c^2-d^2\right)+\hat{i}b\left(-a^2-b^2+c^2+d^2\right)$$
$$+\hat{j}c\left(-a^2-b^2+c^2+d^2\right)+\hat{k}d\left(-a^2-b^2+c^2+d^2\right)$$

with relations: $\left\{\hat{i}^2=-1,\hat{j}^2=+1,\hat{k}^2=+1,\hat{i}\hat{j}=-\hat{j}\hat{i}=\pm\hat{k},\hat{i}\hat{k}=-\hat{k}\hat{i}=\pm\hat{j},\hat{j}\hat{k}=-\hat{k}\hat{j}=\pm\hat{i},\right\}$

Matrix Form 2:

$$\begin{bmatrix} \mathbb{S} & \mathbb{OP}_\mathbb{S} \\ \mathbb{OP}_\mathbb{S} & \mathbb{S} \end{bmatrix} \triangleq \begin{bmatrix} a & b & c & -d \\ b & a & d & -c \\ c & -d & a & b \\ d & -c & b & a \end{bmatrix}$$

This is the algebra of a hyperbolic complex number and an odd parity number set in the matrix form of a hyperbolic complex number. This matrix form is multiplicatively closed but not multiplicatively commutative. It is inversely closed with: $\det = \left(a^2-b^2-c^2+d^2\right)^2$ which is positive definite for all values of $\{a,b,c,d\}$. We take the norm to be the fourth root of the determinant.

In non-matrix notation, the number is: $a+\hat{r}b+\hat{s}c+\hat{t}d$. The conjugate is:

$$a\left(a^2 - b^2 - c^2 + d^2\right) + \hat{r}b\left(-a^2 + b^2 + c^2 - d^2\right)$$
$$+\hat{s}c\left(-a^2 + b^2 + c^2 - d^2\right) + \hat{t}d\left(-a^2 + b^2 + c^2 - d^2\right)$$

with relations: $\left\{\hat{r}^2 = +1, \hat{s}^2 = +1, \hat{t}^2 = -1, \hat{r}\hat{s} = -\hat{s}\hat{r} = \pm\hat{t}, \hat{r}\hat{t} = -\hat{t}\hat{r} = \pm\hat{s}, \hat{s}\hat{t} = -\hat{t}\hat{s} = \pm\hat{r},\right\}$

Matrix form 3: - The quaternion algebra of William Hamilton.

$$\begin{bmatrix} \mathbb{C} & \mathbb{OP}_{\mathbb{C}} \\ -\mathbb{OP}_{\mathbb{C}} & \mathbb{C} \end{bmatrix} \triangleq \begin{bmatrix} a & b & c & d \\ -b & a & d & -c \\ -c & -d & a & b \\ -d & c & -b & a \end{bmatrix}$$

This is the algebra of a euclidean complex number and an odd parity number set in the matrix form of a euclidean complex number. This matrix form is multiplicatively closed but not multiplicatively commutative. It is inversely closed with: $\det = \left(a^2 + b^2 + c^2 + d^2\right)^2$ which is positive definite for all values of $\{a,b,c,d\}$. We take the norm to be the fourth root of the determinant.

In non-matrix notation, the number is: $a + \hat{i}b + \hat{j}c + \hat{k}d$. The conjugate, taken from the adjoint matrix, is:

$$a\left(a^2 + b^2 + c^2 + d^2\right) + \hat{i}b\left(-a^2 - b^2 - c^2 - d^2\right)$$
$$+\hat{j}c\left(-a^2 - b^2 - c^2 - d^2\right) + \hat{k}d\left(-a^2 - b^2 - c^2 - d^2\right)$$

with relations: $\left\{\hat{i}^2 = -1, \hat{j}^2 = -1, \hat{k}^2 = -1, \hat{i}\hat{j} = -\hat{j}\hat{i} = \pm\hat{k}, \hat{i}\hat{k} = -\hat{k}\hat{i} = \pm\hat{j}, \hat{j}\hat{k} = -\hat{k}\hat{j} = \pm\hat{i},\right\}$

Matrix Form 4:

$$\begin{bmatrix} \mathbb{S} & \mathbb{OP}_{\mathbb{S}} \\ -\mathbb{OP}_{\mathbb{S}} & \mathbb{S} \end{bmatrix} \triangleq \begin{bmatrix} a & b & c & -d \\ b & a & d & -c \\ -c & d & a & b \\ -d & c & b & a \end{bmatrix}$$

This is the algebra of a hyperbolic complex number and an odd parity number set in the matrix form of a euclidean complex number. This matrix form is multiplicatively closed but not multiplicatively commutative. It is inversely closed with: $\det = \left(a^2 - b^2 + c^2 - d^2\right)^2$ which is positive definite for all values of $\{a,b,c,d\}$. We take the norm to be the fourth root of the determinant.

In non-matrix notation, the number is: $a + \hat{r}b + \hat{s}c + \hat{t}d$. The conjugate is:

$$a\left(a^2 - b^2 + c^2 - d^2\right) + \hat{r}b\left(-a^2 + b^2 - c^2 + d^2\right)$$
$$+\hat{s}c\left(-a^2 + b^2 - c^2 + d^2\right) + \hat{t}d\left(-a^2 + b^2 - c^2 + d^2\right)$$

with relations: $\left\{\hat{r}^2 = +1, \hat{s}^2 = -1, \hat{t}^2 = +1, \hat{r}\hat{s} = -\hat{s}\hat{r} = \pm\hat{t}, \hat{r}\hat{t} = -\hat{t}\hat{r} = \pm\hat{s}, \hat{s}\hat{t} = -\hat{t}\hat{s} = \pm\hat{r},\right\}$

Connections to the Clifford Algebras

Matrix Form 1 is the hyperbolic complex numbers with variables that are sums of parts of the Cl_2 Clifford algebra:

$$\mathbb{S}_{Cl_2}\left(a+b\overline{e_1}\overline{e_2},c\overline{e_1}+d\overline{e_2}\right)$$

Matrix Form 2 is the same but with elements from the Clifford algebra $Cl_{1,1}$:

$$\mathbb{S}_{Cl_{1,1}}\left(a+b\overline{e_1}\overline{e_2},c\overline{e_1}+d\overline{e_2}\right)$$

Matrix Form 3 is the euclidean complex numbers with variables that are sums of parts of the Cl_2 Clifford algebra:

$$\mathbb{C}_{Cl_2}\left(a+b\overline{e_1}\overline{e_2},c\overline{e_1}+d\overline{e_2}\right)$$

Matrix Form 4 is the same but with elements from the Clifford algebra $Cl_{1,1}$:

$$\mathbb{C}_{Cl_{1,1}}\left(a+b\overline{e_1}\overline{e_2},c\overline{e_1}+d\overline{e_2}\right)$$

Groups Associated with these Matrix Forms:

The unit sub-matrices[37], together with their additive inverses, of the matrix form 3 (William Hamilton's quaternions) form the quaternion group, Q. The unit sub-matrices, with their additive inverses, of each of the other three matrix forms form the dihedral group, D_4. In this sense, there is only one quaternion algebra, and the other three algebras in this chapter are dihedral algebras. We refer to them as dihedralions.

The quaternion group has three C_4 subgroups. The three C_4 subgroups are generated by $\{b,c,d\}$. The dihedral group, D_4, has one C_4 subgroup. In each of these three dihedral algebras, the element that generates the single C_4 subgroup is different. In matrix form 1, the C_4 subgroup is generated by b. In matrix form 2, the C_4 subgroup is generated by d. In matrix form 4, the C_4 subgroup is generated by c. The three dihedralion algebras are algebraically isomorphic. It is the sub-group structure that determines that there are three 4-dimensional dihedral algebras and only one 4-dimensional quaternion algebra.

---

[37] Take an algebraic matrix form, set one variable to unity, set the other variables to zero; you now have a unit sub-matrix.

Section 3 : Chapter 6

The Hyperbolic Complex Numbers are not Algebraically Closed

Any polynomial with real or euclidean complex numbers as coefficients can be factored into linear factors over the euclidean complex numbers. For example, the polynomial $x^2 + 1 = (x + \hat{i})(x - \hat{i}) : \hat{i} = \sqrt{-1}$ cannot be factored into linear factors over the real numbers but it can be factored into linear factors over the euclidean complex numbers. This property, that all polynomials will split into linear factors over the euclidean complex numbers, is called the algebraic closure of the euclidean complex numbers.

Consider the polynomial:

$$\begin{bmatrix} x^2 & 0 \\ 0 & x^2 \end{bmatrix} + \begin{bmatrix} 1 & 0 \\ 0 & 1 \end{bmatrix}$$

If this will split into linear factors then we would have:

$$\left(\begin{bmatrix} x & 0 \\ 0 & x \end{bmatrix} + \begin{bmatrix} a & b \\ b & a \end{bmatrix} \right) \left(\begin{bmatrix} x & 0 \\ 0 & x \end{bmatrix} + \begin{bmatrix} c & d \\ d & c \end{bmatrix} \right) = \begin{bmatrix} x^2 & 0 \\ 0 & x^2 \end{bmatrix} + \begin{bmatrix} 1 & 0 \\ 0 & 1 \end{bmatrix}$$

However:

$$\left(\begin{bmatrix} x & 0 \\ 0 & x \end{bmatrix} + \begin{bmatrix} a & b \\ b & a \end{bmatrix} \right) \left(\begin{bmatrix} x & 0 \\ 0 & x \end{bmatrix} + \begin{bmatrix} c & d \\ d & c \end{bmatrix} \right)$$

$$= \begin{bmatrix} x^2 & 0 \\ 0 & x^2 \end{bmatrix} + \begin{bmatrix} ax & bx \\ bx & ax \end{bmatrix} + \begin{bmatrix} cx & dx \\ dx & cx \end{bmatrix} + \begin{bmatrix} ac + bd & ad + bc \\ ad + bc & ac + bd \end{bmatrix}$$

$$\Rightarrow a = -c, b = -d \Rightarrow$$

$$\left(\begin{bmatrix} x & 0 \\ 0 & x \end{bmatrix} + \begin{bmatrix} a & b \\ b & a \end{bmatrix} \right) \left(\begin{bmatrix} x & 0 \\ 0 & x \end{bmatrix} + \begin{bmatrix} c & d \\ d & c \end{bmatrix} \right) = \begin{bmatrix} x^2 & 0 \\ 0 & x^2 \end{bmatrix} + \begin{bmatrix} -(a^2 + b^2) & -2ab \\ -2ab & -(a^2 + b^2) \end{bmatrix}$$

$$\Rightarrow 2ab = 0, a^2 + b^2 = -1$$

$$\Rightarrow a = 0, b = \sqrt[2]{-1} \text{ or } b = 0, a = \sqrt[2]{-1}$$

Since $\sqrt[2]{-1}$ is not an element of the hyperbolic complex numbers, the hyperbolic complex numbers are not algebraically closed.

Theorem:
The Hyperbolic Complex Numbers are not Algebraically Closed
Proof:
See above

Section 3 : Chapter 7

Vectors within Hyperbolic Space

There is a one-to-one correspondence between the algebra of the hyperbolic complex numbers and 2-dimensional vectors.

$$\begin{bmatrix} a & b \\ b & a \end{bmatrix} \leftrightarrow \begin{bmatrix} a \\ b \end{bmatrix}$$

The angle operation is:

$$\begin{bmatrix} a & b \\ b & a \end{bmatrix} \odot \begin{bmatrix} c & d \\ d & c \end{bmatrix} = \begin{bmatrix} a & b \\ b & a \end{bmatrix} \begin{bmatrix} c & -d \\ -d & c \end{bmatrix}$$

$$= \begin{bmatrix} ac-bd & -(ad-bc) \\ -(ad-bc) & ac-bd \end{bmatrix}$$

Leading to:

$$\cosh \chi = \frac{ac-bd}{\sqrt{a^2-b^2}\sqrt{c^2-d^2}}$$

$$\sinh \chi = \frac{ad-bc}{\sqrt{a^2-b^2}\sqrt{c^2-d^2}}$$

The first of these is the established inner product of Minkowski space-time. Since $\cosh \chi$ is never zero, orthogonality has no meaning in this space.

The second of the above expressions is the magnitude of the wedge product in Minkowski space-time. Other than the normalising terms, the magnitude of the wedge product of Minkowski space-time is the same as the magnitude of the wedge product of 2-dimensional euclidean space. In the same way that the 'imaginary' part of a euclidean complex number is associated with rotation, the 'imaginary' part of a hyperbolic complex number is associated with rotation. Thus, the wedge product in Minkowski space-time is associated with rotation[38].

The differential of a scalar field is:

$$d \begin{bmatrix} f(a,b) & 0 \\ 0 & f(a,b) \end{bmatrix} = \begin{bmatrix} \dfrac{\partial f}{\partial a} & \dfrac{\partial f}{\partial b} \\ \dfrac{\partial f}{\partial b} & \dfrac{\partial f}{\partial a} \end{bmatrix}$$

which is the gradient in 2-dimensional hyperbolic space.

The differential of a vector field is:

---

[38] Setting $\left\{ \dfrac{b}{a} = v, \dfrac{d}{c} = 0 \right\}$ leads to the transformation of special relativity.

$$d\begin{bmatrix} F_a(a,b) & F_b(a,b) \\ F_b(a,b) & F_a(a,b) \end{bmatrix} = \begin{bmatrix} \dfrac{\partial F_a(a,b)}{\partial a} + \dfrac{\partial F_b(a,b)}{\partial b} & \dfrac{\partial F_b(a,b)}{\partial a} + \dfrac{\partial F_a(a,b)}{\partial b} \\ \dfrac{\partial F_b(a,b)}{\partial a} + \dfrac{\partial F_a(a,b)}{\partial b} & \dfrac{\partial F_a(a,b)}{\partial a} + \dfrac{\partial F_b(a,b)}{\partial b} \end{bmatrix}$$

$$= \begin{bmatrix} div(F) & curl(F) \\ curl(F) & div(F) \end{bmatrix}$$

which are the divergence and curl of a vector field in space-time[39].

The Laplacian of a scalar field is:

$$\begin{bmatrix} \dfrac{\partial^2 f}{\partial a^2} + \dfrac{\partial^2 f}{\partial b^2} & 2\dfrac{\partial^2 f}{\partial a \partial b} \\ 2\dfrac{\partial^2 f}{\partial a \partial b} & \dfrac{\partial^2 f}{\partial a^2} + \dfrac{\partial^2 f}{\partial b^2} \end{bmatrix}$$

which, in this space, is a vector.

The Laplacian of a vector field is:

$$\begin{bmatrix} \dfrac{\partial^2 F_a}{\partial a^2} + \dfrac{\partial^2 F_a}{\partial b^2} + 2\dfrac{\partial^2 F_b}{\partial a \partial b} & \dfrac{\partial^2 F_b}{\partial a^2} + \dfrac{\partial^2 F_b}{\partial b^2} + 2\dfrac{\partial^2 F_a}{\partial a \partial b} \\ \dfrac{\partial^2 F_b}{\partial a^2} + \dfrac{\partial^2 F_b}{\partial b^2} + 2\dfrac{\partial^2 F_a}{\partial a \partial b} & \dfrac{\partial^2 F_a}{\partial a^2} + \dfrac{\partial^2 F_a}{\partial b^2} + 2\dfrac{\partial^2 F_b}{\partial a \partial b} \end{bmatrix}$$

---

[39] Hm!

Introduction to Section 4

Within this section we meet 1-dimensional space and see how it is conventionally used to construct higher dimensional spaces. We also introduce the reader to the general 2-dimensional rotation matrix.

Section 4 : Chapter 1

1-dimensional Space

The algebra of the real numbers is associated with 1-dimensional space. The real numbers have matrix form that is a 1×1 matrix. We call 1-dimensional space the 'real number line'. There is only one 1-dimensional algebra – the real numbers, and so there is only one 1-dimensional space. This space is homogeneous in that the real numbers are the same distance apart at every position within this space. Isotropy is a meaningless concept in 1-dimensional space. There is no concept of angle in 1-dimensional space.

The polar form of the real numbers is:

$$\exp\big([a]\big) = \exp\big([a] + [0]\big)$$
$$= \exp\big([a]\big)\exp\big([0]\big)$$
$$= \big[e^{a}\big][1]$$

$[1]$ is the 'rotation matrix' of 1-dimensional space. We jump the gun a little, bend the terminology a little, and declare that this rotation matrix contains the 1-dimensional trigonometric function, unity. This function takes zero arguments.

Mathematicians are wont to construct higher dimensional spaces by fixing together a number of 1-dimensional spaces; we will state the procedure.
a) Take n 1-dimensional spaces.
b) Fix them together at the point zero, which will hereafter be called the origin.
c) Within these 1-dimensional spaces, there is no concept of n-dimensional distance. Such a concept is needed, and so the mathematician invents it. The mathematician will not accept any old n-dimensional distance function; it is required to satisfy various axioms (usually the metric space axioms), but, none-the-less, the function comes from within the mathematician's head not from within the algebra.
c) There is no concept of angle within 1-dimensional space. Such a concept is needed, and so the mathematician invents it. It is called the inner product. Often, but not necessarily, the inner product is partly derived from the invented distance function. The inner product cannot be wholly derived from the distance function because its construction requires a trigonometric function of some kind that accepts an argument – an angle. The required trigonometric function is taken from a 2-dimensional space. Usually, it is the cos trigonometric function that is taken, but the cosh function is taken sometimes. And there we have it:

> Conventional *n*-dimensional space:
> > *n* 1-dimensional spaces
> > An invented distance function
> > An invented inner-product
> > A kidnapped trigonometric function

Such is conventional *n*-dimensional space. The natural spaces are different.

Section 4 : Chapter 2

The General Rotation Matrix – 2-dimensions

The general form of the 2-dimensional algebraic matrix form is:

$$\begin{bmatrix} a & b \\ jb & a \end{bmatrix} : j \neq 0$$

We have:

$$\exp\left(\begin{bmatrix} a & b \\ jb & a \end{bmatrix}\right) = \begin{bmatrix} e^a & 0 \\ 0 & e^a \end{bmatrix} \begin{bmatrix} \cosh\left(\sqrt{jb}\right) & \frac{1}{\sqrt{j}}\sinh\left(\sqrt{jb}\right) \\ j\frac{1}{\sqrt{j}}\sinh\left(\sqrt{jb}\right) & \cosh\left(\sqrt{jb}\right) \end{bmatrix}$$

This is the rotation matrix of the 2-dimensional natural space. When $j = -1$, this is the euclidean complex numbers. Thus, we can write

$$\begin{bmatrix} \cosh\left(\hat{i}b\right) & -\hat{i}\sinh\left(\hat{i}b\right) \\ \hat{i}\sinh\left(\hat{i}b\right) & \cosh\left(\hat{i}b\right) \end{bmatrix} = \begin{bmatrix} \cos b & \sin b \\ -\sin b & \cos b \end{bmatrix}$$

This implies:

$$\cosh\left(\hat{i}b\right) = \cos\left(b\right)$$

$$\sinh\left(\hat{i}b\right) = \hat{i}\sin\left(b\right)$$

Introduction to Section 5

In previous sections, we have met two 2-dimensional algebraic matrix forms (the euclidean complex numbers and the hyperbolic complex numbers) and one 1-dimensional algebraic matrix form (the real numbers). Are there any other algebraic matrix forms?

In this section, we unveil the matrix forms of the other natural algebraic fields. We do this by deducing the properties of algebraic matrix forms from the algebraic field axioms. We also look at 'illegitimate' algebraic matrix forms that are formed by putting two or more algebraic matrix forms as blocks onto the leading diagonal of an appropriately sized matrix.

Section 5 : Chapter 1

The Algebraic Matrix Forms from the Field Axioms

The axioms of an algebraic field are given in Appendix 2. In this work, we are considering only sets of matrices with matrix addition and matrix multiplication. Since any set of matrices with matrix addition and matrix multiplication automatically satisfies many of the axioms of an algebraic field, we do not need to consider every field axiom. We have to consider only:

The Axioms of an Algebraic Matrix Field

F3: Inclusion of the additive identity, $\begin{bmatrix} 0 & 0 \\ 0 & 0 \end{bmatrix}$[40].

F4: Inclusion of additive inverses
F5: Additive Closure
F6: Multiplicative Closure
F7: Inclusion of multiplicative inverses

F8: Inclusion of the multiplicative identity, $\begin{bmatrix} 1 & 0 \\ 0 & 1 \end{bmatrix}$.

F11: Multiplication is commutative
F14: There are no singular matrices in the set

The Characterization of Matrix Forms:

A matrix form may be such that all of its elements are independent of each other or some of its elements are dependent upon others of its elements. We characterize such a matrix form by specifying the relationships between the elements or by writing the matrix; for example:

$$\{a_{11} \in \mathbb{R}, a_{12} \in \mathbb{R}, a_{21} = -a_{12}, a_{22} = 3\} \equiv \begin{bmatrix} a & b \\ -b & 3 \end{bmatrix}$$

---

[40] We demonstrate with 2×2 matrices, but the same applies to larger matrices

Theorem 1
Algebraic matrix forms must be square matrices.
Proof:
The requirement for multiplicative commutativity.

The required inclusion of the multiplicative identity together with the requirement for additive closure allows us to build scalars.

$$\begin{bmatrix} 1 & 0 \\ 0 & 1 \end{bmatrix} + \begin{bmatrix} 1 & 0 \\ 0 & 1 \end{bmatrix} = \begin{bmatrix} 2 & 0 \\ 0 & 2 \end{bmatrix}$$

Multiplicative closure[41] ensures that multiplication by a scalar is inclusive.

Theorem 2
The elements of algebraic matrix forms cannot be (or contain) non-zero constants.
Proof:
The requirement for multiplicative closure after multiplication by a scalar.

Theorem 3
The elements of algebraic matrix forms must be either independent of other elements or linear functions of the other elements.
Proof:
The requirement for multiplicative closure after multiplication by a scalar.

We demonstrate what is stated in Theorem 3 is necessary to maintain the form of the matrix.

$$\begin{bmatrix} 2 & 0 \\ 0 & 2 \end{bmatrix}\begin{bmatrix} a & 0 \\ 0 & a^2 \end{bmatrix} = \begin{bmatrix} 2a & 0 \\ 0 & 2a^2 \end{bmatrix}$$

Theorem 3 means that all the elements of an algebraic matrix form are either independent, $a_{ij} \in \mathbb{R}$, of other elements or are of the form:

$$a_{st} = ja_{zy} + ka_{xw} + ... : \{j, k, ...\} \in \mathbb{R}$$

Theorem 4
The elements on the leading diagonal[42] of an algebraic matrix form must be independent of all elements that are not on the leading diagonal.
Proof:
The requirement for inclusion of the multiplicative identity and Theorem 3.
Theorem 5
The elements on the leading diagonal of an algebraic matrix form are all dependent upon each other. (From Theorem 3, the dependence is linear.)
Proof:

---

[41] Additive/Multiplicative closure means that the sum/product matrix must be of the same form as the matrices from which it is made.
[42] The leading diagonal of a square matrix is the diagonal that runs from the top left-hand corner to the bottom right-hand corner.

We require guaranteed non-singularity for every non-zero matrix. In particular we require non-singularity when all variables within a matrix are zero except one. If the leading diagonal has different independent variables upon it, the requirement for guaranteed non-singularity cannot be met.

Although theorem 5 is valid, we could avoid non-singularity by imposing restrictions that disallow any leading diagonal element from being zero unless all such elements are zero.[43]

Theorem 6
The elements on the leading diagonal of an algebraic matrix form are all equal.
Proof:
Theorem 3 with Theorem 5 and the requirement to include the multiplicative identity.

Corollary:
Every algebraic matrix form contains, as a sub-algebra, an algebra that is isomorphic to the real numbers.
Proof:
Matrices that have all elements of the leading diagonal equal and all other element zero are algebraically isomorphic to the real numbers.

Theorem 7
The only dependency relation is of the form $a_{st} = ja_{xy} : j \in \mathbb{R}$
Proof:
The requirement of the inclusion of the multiplicative inverse and that the elements on the leading diagonal of the multiplicative inverse must be equal – Theorem 6 – and independent of the other elements – Theorem 4.

Theorem 7 over-reaches Theorem 3, which is now redundant.

Theorem 8
For an $n \times n$ matrix with n independent variables, each variable or a multiple of each variable in an algebraic matrix form must occur once and only once in each column and in each row of the matrix.
Proof:
A matrix form can be an algebra only if the matrix form is guaranteed non-singular. This guarantee needs to apply when all but one of the variables within the matrix form is zero. A matrix with a row or column of zeros is always singular. The 'only once' is because there are only n places in each row/column and we need to fit n variables into the row/column.

---

[43] This is exactly what we will do when we consider matrices of the form: $\begin{bmatrix} a & b & 0 \\ -b & a & 0 \\ 0 & 0 & c \end{bmatrix}$

Alternative Proof:
This is required in order to maintain equality of the elements on the leading diagonal under multiplication.

Theorem 8 deals with $n \times n$ matrices with n independent variables. $n \times n$ matrices with less than n independent variables can be deduced from the n independent variables matrices by simply tying two or more variables together. We have no interest in such matrices. Theorem 8 is very restrictive. It is sufficiently restrictive to serve our needs.

We are now in a position to begin forming prospective algebraic matrix forms. We are guided by:

    a) The leading diagonal elements are all equal – Theorem 6

    b) A $n \times n$ algebraic matrix form has n independent variables, and these variables each occur once and only once in every row and in every column of the algebraic matrix form - Theorem 8.

    c) The only dependency relation is of the form $a_{st} = ja_{xy} : j \in \mathbb{R}$ - Theorem 7

How to Construct Algebraic Matrix Forms
1) Choose the size of the $n \times n$ square matrix.
2) Fill the leading diagonal with a's.

$$\begin{bmatrix} a & \\ & a \end{bmatrix}$$

3) Find every possible matrix with $(n-1)$ other independent variables arranged so that each variable occurs once and only once in every column and in every row of the matrix.

$$\begin{bmatrix} a & b \\ b & a \end{bmatrix}$$

4) Other than the leading diagonal and the top row, multiply the other elements of the matrix by a (space squeezing) parameter.

$$\begin{bmatrix} a & b \\ jb & a \end{bmatrix}$$

5) Multiply the matrix thus formed by itself and adjust the parameters to ensure the form of the matrix is maintained by multiplication.

$$\begin{bmatrix} a & b \\ jb & a \end{bmatrix}^2 = \begin{bmatrix} a^2 + jb^2 & 2ab \\ 2jab & a^2 + jb^2 \end{bmatrix}$$

(The case of the 3×3 matrices is below.)

Nomenclature:
Space Squeezing Parameters
The parameters that occur in algebraic matrix forms that determine the dependency of one element upon another are called space squeezing parameters.

Having so formed prospective algebraic matrix forms in accordance with the above, we will have to check the algebraic matrix field axioms to ensure that they do form an algebraic field.

F3 is automatically satisfied for matrices whose elements are unrestricted real variables that are either independent or are linear multiples of each other.

F4 is automatically satisfied for matrices whose elements are unrestricted real variables.

F5 is automatically satisfied for matrices whose elements are unrestricted real variables that are either independent or are linear multiples of each other.

F6 is automatically satisfied for square matrices whose elements are independent real variables that appear once and only once in each row and in each column.

F7 is automatically satisfied for square matrices whose elements are independent real variables that appear once and only once in each row and in each column provided F14 is also satisfied.

F8 is automatically satisfied for matrices whose leading diagonal elements are all equal real variables that are independent of all other elements

This leaves us with only axioms F11 & F14 to check.

F11: There are matrices of the form we have constructed that are not multiplicatively commutative; the quaternion matrix form is one example.

F14: There are matrices of the form we have constructed whose cartesian forms are not guaranteed non-singular.

Theorem 9[44]

Multiplicatively commutative matrices of the form we have constructed are guaranteed non-singular in their polar form.

Proof:

Since the matrices are multiplicatively commutative for all values of the variables, then they are multiplicatively commutative when all variables except one are zero. We can therefore split the exponentiated form of these matrices into exponentiated separate variable sub-matrices. One of these sub-matrices will be the sub-matrix with the leading diagonal variable; the other sub-matrices will have trace equal to zero. Thus the determinant of the polar form of the matrix will never be singular.

We have seen theorem 9 in action when we took the polar form of the hyperbolic complex numbers. We now demonstrate theorem 9 in a 3-dimensional case:

$$\exp\left(\begin{bmatrix} a & b & c \\ c & a & b \\ b & c & a \end{bmatrix}\right) = \exp\left(\begin{bmatrix} a & 0 & 0 \\ 0 & a & 0 \\ 0 & 0 & a \end{bmatrix} + \begin{bmatrix} 0 & b & 0 \\ 0 & 0 & b \\ b & 0 & 0 \end{bmatrix} + \begin{bmatrix} 0 & 0 & c \\ c & 0 & 0 \\ 0 & c & 0 \end{bmatrix}\right)$$

$$= \exp\left(\begin{bmatrix} a & 0 & 0 \\ 0 & a & 0 \\ 0 & 0 & a \end{bmatrix}\right)\exp\left(\begin{bmatrix} 0 & b & 0 \\ 0 & 0 & b \\ b & 0 & 0 \end{bmatrix}\right)\exp\left(\begin{bmatrix} 0 & 0 & c \\ c & 0 & 0 \\ 0 & c & 0 \end{bmatrix}\right)$$

---

[44] If this theorem sounds confusing now, don't worry. It will become crystal clear later in the work.

$$\det\left(\exp\left(\begin{bmatrix} a & b & c \\ c & a & b \\ b & c & a \end{bmatrix}\right)\right) = PROD\left(\begin{array}{l} \det\left(\begin{bmatrix} e^a & 0 & 0 \\ 0 & e^a & 0 \\ 0 & 0 & e^a \end{bmatrix}\right) \\ \det\left(\exp\left(\begin{bmatrix} 0 & b & 0 \\ 0 & 0 & b \\ b & 0 & 0 \end{bmatrix}\right)\right) \det\left(\exp\left(\begin{bmatrix} 0 & 0 & c \\ c & 0 & 0 \\ 0 & c & 0 \end{bmatrix}\right)\right) \end{array}\right)$$

$$= e^{3a}.\exp\left(Trace\left(\begin{bmatrix} 0 & b & 0 \\ 0 & 0 & b \\ b & 0 & 0 \end{bmatrix}\right)\right)\exp\left(Trace\left(\begin{bmatrix} 0 & 0 & c \\ c & 0 & 0 \\ 0 & c & 0 \end{bmatrix}\right)\right)$$

$$= e^{3a}.1.1 \neq 0$$

Thus, if we use the polar forms (or only matrices equivalent to a polar form), we can ignore F14.[45] We need to consider only F11 - multiplicative commutativity. This is what we will do.

Important Point:
In each case, the set of matrices that form the algebra is the set of matrices that are equal to the polar form matrices of the algebra. In general, this is not all of the set of the cartesian matrices. In the case of the euclidean complex numbers, it is the whole cartesian set. In the case of the hyperbolic complex numbers, it is a restricted part of the cartesian set. Thus, by using only the polar form matrices, we avoid having to impose restrictions upon the values that the variables within the algebra can take.

Theorem 10
Matrices constructed as detailed in 'How to Construct Algebraic Matrix Forms' above are algebraic fields if they are multiplicatively commutative.
Proof:
Given above

An Aside:
    If such matrices are not multiplicatively commutative, we cannot form the polar form and thereby be rid of the requirement for non-singularity. However, such matrices will form a multiplicatively non-commutative division algebra with appropriate restrictions applied to their cartesian form. The quaternions are atypical in that they require no restrictions.

The 1×1 Matrices:
The 1×1 matrices are the real numbers, and these are an algebraic field.

---

[45] Which is one huge headache out of the way! However, the physical constant that is the limiting velocity in space-time is derived from the restrictions. Perhaps other physical constants lurk in the restrictions of other algebras.

The 2×2 Matrices:
The only prospective form is:

$$\begin{bmatrix} a & b \\ jb & a \end{bmatrix} : j \in \mathbb{R}$$

We need to impose the restriction $|a| > \left|\sqrt{jb}\right|$ to make this into an algebraic field and thus an algebraic matrix form.

Most often, we will be concerned only with the cases $j = \pm 1$ (and similarly for other such parameters). We call such algebras flat algebras.

Nomenclature:
Flat Algebras or Flat Algebraic Fields
An algebraic matrix form with spaces squeezing parameters equal to plus or minus unity.

When $j = -1$, we have:

$$\mathbb{C} \cong \begin{bmatrix} a & b \\ -b & a \end{bmatrix}$$

This is the euclidean complex numbers in matrix notation. The case $j = 1$ gives:

$$\mathbb{S} \cong \begin{bmatrix} a & b \\ b & a \end{bmatrix} : |a| > |b|$$

This is the hyperbolic complex numbers.

We put all this into a theorem.

Theorem 11
The only forms of 2×2 matrices that are an algebraic field are[46]:

$$\left\{ \begin{bmatrix} a & b \\ jb & a \end{bmatrix} : j < 0; \begin{bmatrix} a & b \\ 0 & a \end{bmatrix} : a \neq 0; \begin{bmatrix} a & b \\ jb & a \end{bmatrix} : |a| > |b|, j > 0 \right\}$$

Proof:
Given above

Corollary: With $j = \pm 1$, the only forms of 2×2 matrices that are a flat algebraic field are:

$$\left\{ \mathbb{C} \cong \begin{bmatrix} a & b \\ -b & a \end{bmatrix}, \mathbb{S} \cong \begin{bmatrix} a & b \\ b & a \end{bmatrix} : |a| > |b| \right\}$$

The 3×3 Matrices:
The only prospective algebraic matrix form is:

---

[46] We count the real numbers as a 1×1 matrix.

$$\begin{bmatrix} a & b & c \\ jc & a & kb \\ lb & mc & a \end{bmatrix}$$

We now use the requirement of equality of leading diagonal elements together with the requirement of multiplicative closure. We multiply this matrix by itself and equate the leading diagonal elements of the product[47].

$$\begin{bmatrix} a & b & c \\ jc & a & kb \\ lb & mc & a \end{bmatrix}^2 = \begin{bmatrix} a^2 + jbc + lbc & \sim & \sim \\ \sim & a^2 + jbc + kmbc & \sim \\ \sim & \sim & a^2 + kmbc + lbc \end{bmatrix}$$

From this we get:

$$\begin{bmatrix} a & b & c \\ jc & a & kb \\ jb & \dfrac{j}{k}c & a \end{bmatrix} : j \in \mathbb{R}, k \in \mathbb{R} \neq 0$$

Simple multiplication shows this form to be multiplicatively commutative, and so this is (in cartesian form with restrictions[48]) an algebraic field. Although this is an algebraic field for the stated values of $\{j,k\}$, our only concern at present is with the flat algebras.

These are given by the four permutations of $\{j,k\} = \pm 1$.

Theorem 12

The only forms of 3×3 matrices with three independent variables that are algebraic fields are (with appropriate restrictions):

$$C_3 = \begin{bmatrix} a & b & c \\ jc & a & kb \\ jb & \left(\dfrac{j}{k}\right)c & a \end{bmatrix} : j \in \mathbb{R}, k \in \mathbb{R} \neq 0$$

Proof:
Given above.

---

[47] With larger matrices, it is necessary to consider the off-diagonal elements also.
[48] We do not dwell upon the restrictions; they are of no importance because the polar forms do not need them. They simply reflect the difficulty of trying to make non-euclidean spaces match the extent of euclidean space.

Corollary: With $\{j,k\}=\pm1$, the only flat forms of 3×3 matrices (with three independent variables) that are an algebraic field are (with restrictions):

$$\left\{ \begin{array}{ll} C_3L^jH^2{}_{(j=1,k=1)} = \begin{bmatrix} a & b & c \\ c & a & b \\ b & c & a \end{bmatrix}, & C_3L^jE^jH^j{}_{(j=-1,k=1)} = \begin{bmatrix} a & b & c \\ -c & a & b \\ -b & -c & a \end{bmatrix} \\[6ex] C_3L^jE^2{}_{(j=1,k=-1)} = \begin{bmatrix} a & b & c \\ c & a & -b \\ b & -c & a \end{bmatrix}, & C_3L^jE^jH^j{}_{(j=-1,k=-1)} = \begin{bmatrix} a & b & c \\ -c & a & -b \\ -b & c & a \end{bmatrix} \end{array} \right\}$$

The names we have appelled to these algebras will seem strange. We provide a pre-emptive explanation below

The Origin of the Names Appelled to these Algebras

The name of an algebra has three components.
i) The name of the group that underlies the algebra. This will mean nothing to the reader yet – it is explained later. If there is more than one algebra derived from a particular group (because sub-groups are generated by different elements of the group), then these different forms are denoted by a decimal point and a number added to the name of the group. Eg: $C_{4.1}$

ii) The polar form of the algebra (if known). Eg: L^jH^3. The L is the (always present) length matrix. The H stands for a rotation matrix with H-type simple trigonometric functions within it. The simple trigonometric functions will be introduced later. The E stands for a rotation matrix with E-type simple trigonometric functions within it.
iii) The subscripted values of the space squeezing parameters that generate the flat form of the algebra. Eg: $(j=1,k=-1)$

4×4 Matrices:
The prospective algebraic matrix forms (with four independent variables) are:

$$\left\{ \begin{array}{ll} \begin{bmatrix} a & b & c & d \\ jb & a & kd & lc \\ mc & nd & a & pb \\ qd & rc & sb & a \end{bmatrix} & \begin{bmatrix} a & b & c & d \\ jb & a & kd & lc \\ md & nc & a & pb \\ qc & rd & sb & a \end{bmatrix} \\[8ex] \begin{bmatrix} a & b & c & d \\ jd & a & kb & lc \\ mc & nd & a & pb \\ qb & rc & sd & a \end{bmatrix} & \begin{bmatrix} a & b & c & d \\ jc & a & kd & lb \\ mb & nd & a & pc \\ qd & rc & sb & a \end{bmatrix} \end{array} \right\}$$

In each case, we multiply this matrix by itself and equate the leading diagonal elements of the product (and a bit more).[49] This leads to:

$$C_{4.1} = \begin{bmatrix} a & b & c & d \\ jd & a & kb & lc \\ \dfrac{jl}{k}c & \dfrac{j}{k}d & a & lb \\ jb & \dfrac{j}{k}c & \dfrac{j}{l}d & a \end{bmatrix} : \{k,l\} \neq 0, \qquad C_{4.2} = \begin{bmatrix} a & b & c & d \\ jb & a & kd & \dfrac{j}{k}c \\ ld & \dfrac{l}{k}c & a & \dfrac{j}{k}b \\ lc & \dfrac{kl}{j}d & kb & a \end{bmatrix} : \{j,k\} \neq 0$$

$$C_{4.3} = \begin{bmatrix} a & b & c & d \\ jc & a & kd & lb \\ jb & \dfrac{j}{l}d & a & \dfrac{j}{k}c \\ \dfrac{jk}{l}d & \dfrac{j}{l}c & kb & a \end{bmatrix} : \{k,l\} \neq 0, \quad C_2 \times C_2 = \begin{bmatrix} a & b & c & d \\ jb & a & \dfrac{j}{l}d & lc \\ kc & \dfrac{k}{l}d & a & lb \\ \dfrac{jk}{l^2}d & \dfrac{k}{l}c & \dfrac{j}{l}b & a \end{bmatrix} : \{l\} \neq 0$$

We check for multiplicative commutativity and find that (with restrictions in the cartesian form) all these matrix forms, with all permutations of $\{j,k,l\} = \{+1,-1\}$, are flat algebraic fields. The first three, $\{C_{4 \cdot 1 \ldots 3}\}$, are algebraically isomorphic to each other[50]. The fourth is a distinct algebra – it is two algebras fixed together in a groupy[51] sort of way.

The above given algebraic matrix forms are the only flat $\{1 \times 1, 2 \times 2, 3 \times 3, 4 \times 4\}$ matrix forms (with n independent variables) that are algebraic fields. Thus, these are the only 'small' natural algebras.

Section 5 : Addendum 1 to Chapter 1

Block Algebras

Because of the block multiplication nature of matrices, larger matrices that incorporate algebraic matrix forms as blocks upon the leading diagonal have many of the

---

[49] When calculating larger algebraic matrix forms, setting the leading diagonal elements equal is insufficient to complete the calculation and we need to look at the structure of the whole matrix. If we were brave, we could exponentiate the general form sub-matrices off the leading diagonal, take the determinants of those exponentiated matrices, and set parameters in a way that cancels the parameters from within those determinants.

[50] Proof is given as an addendum to this chapter.

[51] Precise and clear prose is indicative of a writer who presumes the reader is of insufficient intelligence to understand sloppy prose. One can combine algebras in a non-groupy way – put them on the leading diagonal.

properties of an algebra. They are most often not a bona-fide algebraic matrix form because the leading diagonal elements are not kept equal; there are exceptions. We present some examples:

$$\begin{bmatrix} a & b & 0 & 0 \\ -b & a & 0 & 0 \\ 0 & 0 & a & b \\ 0 & 0 & b & a \end{bmatrix}^2 = \begin{bmatrix} a^2-b^2 & 2ab & 0 & 0 \\ -2ab & a^2-b^2 & 0 & 0 \\ 0 & 0 & a^2+b^2 & 2ab \\ 0 & 0 & 2ab & a^2+b^2 \end{bmatrix}$$

$$\begin{bmatrix} a & b & c & 0 \\ c & a & b & 0 \\ b & c & a & 0 \\ 0 & 0 & 0 & 1 \end{bmatrix}^2 = \begin{bmatrix} a^2+2bc & c^2+2ab & b^2+2ac & 0 \\ b^2+2ac & a^2+2bc & c^2+2ab & 0 \\ c^2+2ab & b^2+2ac & a^2+2bc & 0 \\ 0 & 0 & 0 & 1 \end{bmatrix}$$

Section 5 : Addendum 2 to Chapter 1

<div align="center">Proof of the Algebraic Isomorphism of the three 4-dimension C_4 type Algebras</div>

Note: In these proofs, we use the non-matrix form of notation.[52] The multiplication operations can be found by inspection of the product matrix of two matrices of a particular algebra.

> **Theorem:**
> $C_{4.1}$ is isomorphic as an algebra to $C_{4.2}$.

Proof:

$$\phi\left(C_{4.1} \mapsto C_{4.2}\right) : \left(a+b\hat{r}+c\hat{s}+d\hat{t} \mapsto a+c\hat{r}+b\hat{s}+b\hat{t}\right)$$

Addition:

$$\phi\begin{pmatrix} \left(a+b\hat{r}+c\hat{s}+d\hat{t}\right)+\left(e+f\hat{r}+g\hat{s}+h\hat{t}\right) \\ =(a+e)+(b+f)\hat{r}+(c+g)\hat{s}+(d+h)\hat{t} \end{pmatrix} \mapsto \begin{pmatrix} (a+e)+(c+g)\hat{r}+(b+f)\hat{s}+(d+h)\hat{t} \\ =\left(a+c\hat{r}+b\hat{s}+d\hat{t}\right)+\left(e+g\hat{r}+f\hat{s}+h\hat{t}\right) \end{pmatrix}$$

The Multiplication:
The multiplication operations within these two algebras are different.

---

[52] The non-matrix form of notation is yet to be dealt with. The reader might like to return to this addendum on a second reading.

$$\phi\begin{pmatrix}\left(a+b\hat{r}+c\hat{s}+d\hat{t}\right)\times C_{4.1}\times\left(e+f\hat{r}+g\hat{s}+h\hat{t}\right)\\=\left(ae+bh+cg+df\right)+\left(af+be+ch+dg\right)\hat{r}+\left(ag+bf+ce+dh\right)\hat{s}+\left(ah+bg+cf+de\right)\hat{t}\end{pmatrix}$$

$$\mapsto\begin{pmatrix}\left(ae+bh+cg+df\right)+\left(ag+bf+ce+dh\right)\hat{r}+\left(af+be+ch+dg\right)\hat{s}+\left(ah+bg+cf+de\right)\hat{t}\\=\left(a+c\hat{r}+b\hat{s}+d\hat{t}\right)\times C_{4.2}\times\left(e+g\hat{r}+f\hat{s}+h\hat{t}\right)\end{pmatrix}$$

Where we have used $\times C_{4.1}\times$ and $\times C_{4.2}\times$ to represent the different multiplication operations within these two algebras – one needs care in doing this.

Multiplication by a scalar:

$$\phi\left(\alpha a+\alpha b\hat{r}+\alpha c\hat{s}+\alpha d\hat{t}\right)\mapsto\left(\alpha a+\alpha d\hat{r}+\alpha c\hat{s}+\alpha b\hat{t}\right)$$

Theorem:

$C_{4.3}$ is isomorphic as an algebra to $C_{4.2}$.

Proof:

$$\phi\left(C_{4.3}\mapsto C_{4.2}\right):\left(a+b\hat{r}+c\hat{s}+d\hat{t}\mapsto a+d\hat{r}+c\hat{s}+b\hat{t}\right)$$

Addition:

$$\phi\begin{pmatrix}\left(a+b\hat{r}+c\hat{s}+d\hat{t}\right)+\left(e+f\hat{r}+g\hat{s}+h\hat{t}\right)\\=\left(a+e\right)+\left(b+f\right)\hat{r}+\left(c+g\right)\hat{s}+\left(d+h\right)\hat{t}\end{pmatrix}\mapsto\begin{pmatrix}\left(a+e\right)+\left(d+h\right)\hat{r}+\left(c+g\right)\hat{s}+\left(b+f\right)\hat{t}\\=\left(a+d\hat{r}+c\hat{s}+b\hat{t}\right)+\left(e+h\hat{r}+g\hat{s}+f\hat{t}\right)\end{pmatrix}$$

The Multiplication:

The multiplication operations within these two algebras are different.

$$\phi\begin{pmatrix}\left(a+b\hat{r}+c\hat{s}+d\hat{t}\right)\times C_{4.3}\times\left(e+f\hat{r}+g\hat{s}+h\hat{t}\right)\\=\left(ae+bg+cf+dh\right)+\left(af+be+ch+dg\right)\hat{r}+\left(ag+bh+ce+df\right)\hat{s}+\left(ah+bf+cg+de\right)\hat{t}\end{pmatrix}$$

$$\mapsto\begin{pmatrix}\left(ae+dg+ch+bf\right)+\left(ah+bg+cf+de\right)\hat{r}+\left(ag+df+ce+bh\right)\hat{s}+\left(af+dh+cg+be\right)\hat{t}\\=\left(a+d\hat{r}+c\hat{s}+b\hat{t}\right)\times C_{4.2}\times\left(e+h\hat{r}+g\hat{s}+f\hat{t}\right)\end{pmatrix}$$

Where we have used $\times C_{4.3}\times$ and $\times C_{4.2}\times$ to represent the different multiplication operations within these two algebras.

Multiplication by a scalar:

$$\phi\left(\alpha a+\alpha b\hat{r}+\alpha c\hat{s}+\alpha d\hat{t}\right)\mapsto\left(\alpha a+\alpha d\hat{r}+\alpha c\hat{s}+\alpha b\hat{t}\right)$$

Theorem:

$C_{4.1}$ is isomorphic as an algebra to $C_{4.3}$.

Proof: Isomorphism is transitive.

Section 5 : Chapter 2

Algebraic Matrix Forms by Cayley Tables

Nomenclature:
Standard Form Cayley Table
We say that the Cayley table of a group is in Standard Form if it is arranged so the identities appear on the leading diagonal. This is achieved by making the left-most column the ordered inverses of the top-most row.[53]

Consider the Standard Form Cayley Table of the groups $\{C_1, C_2, C_3, C_4, C_2 \times C_2\}$. We list those Standard Form Cayley Tables alongside the Algebraic Matrix Forms $\{C_1, C_2, C_3, C_{4.1}, C_2 \times C_2\}$ within which the space squeezing parameters have been set to unity:

$$A \qquad : \qquad [a]$$

$$\begin{matrix} A & B \\ B & A \end{matrix} \qquad : \qquad \begin{bmatrix} a & b \\ b & a \end{bmatrix}$$

$$\begin{matrix} A & B & C \\ C & A & B \\ B & C & A \end{matrix} \qquad : \qquad \begin{bmatrix} a & b & c \\ c & a & b \\ b & c & a \end{bmatrix}$$

$$\begin{matrix} A & B & C & D \\ D & A & B & C \\ C & D & A & B \\ B & C & D & A \end{matrix} \qquad : \qquad \begin{bmatrix} a & b & c & d \\ d & a & b & c \\ c & d & a & b \\ b & c & d & a \end{bmatrix}$$

$$\begin{matrix} A & B & C & D \\ B & A & D & C \\ C & D & A & B \\ D & C & B & A \end{matrix} \qquad : \qquad \begin{bmatrix} a & b & c & d \\ b & a & d & c \\ c & d & a & b \\ d & c & b & a \end{bmatrix}$$

These algebraic matrix forms can be constructed by simply copying the Standard Form Cayley Tables into a matrix. Of course, one then has to calculate the space squeezing parameters.

---

[53] The Standard Form of a Cayley table is not unique.

In the previous chapter, we discovered how to construct algebraic matrix forms. The first three steps of that construction are:

1) Choose the size of the $n \times n$ square matrix.
2) Fill the leading diagonal with a's.
3) Find every possible matrix with $(n-1)$ other independent variables arranged so that each variable occurs once and only once in every column and in every row of the matrix.

Cayley tables in standard form are always square and of a size that is the order of the group. Cayley tables in standard form always have the same element (the identity) on the leading diagonal. Cayley tables in any form always have each element of the group occurring once and only once in each row and in each column of the table.

Of course, we still have to check for multiplicative commutativity, and the cartesian form of the algebraic matrix form will require restrictions on the values of the variables.

Conjecture:
It seems to be the case that abelian groups always produce multiplicatively commutative algebraic matrix forms and that non-abelian groups always produce multiplicatively non-commutative algebraic matrix forms – division algebras. The conjecture is that this is the case for all finite groups.

Section 5 : Chapter 3

Illegitimate Spaces

In the first chapter of this section, we chose that all the elements on the leading diagonal must be equal[54] rather than impose restrictions upon the values of the variables. In this chapter, we look at a matrix form that has different variables upon the leading diagonal. We do this because this particular matrix form is supposedly the matrix form of the 3-dimensional space we see about us.

Consider:

$$\begin{bmatrix} a & b & 0 \\ -b & a & 0 \\ 0 & 0 & z \end{bmatrix}$$

These matrices are:
   Inclusive of the additive identity
   Inclusive of additive inverses
   Additively closed
   Multiplicatively closed
   Inclusive of multiplicative inverses
   Inclusive of multiplicative identity
   Multiplicatively commutative
   Non-singular if $z \neq 0$ unless $\{a,b\} = 0$

With the restrictions, these matrices are an algebraic field.

We split this into two parts:

$$\begin{bmatrix} a & 0 & 0 \\ 0 & a & 0 \\ 0 & 0 & z \end{bmatrix} + \begin{bmatrix} 0 & b & 0 \\ -b & 0 & 0 \\ 0 & 0 & 0 \end{bmatrix}$$

We note that these two parts are multiplicatively commutative, and so we can exponentiate this matrix and split it.
We have:

$$\exp\left(\begin{bmatrix} a & 0 & 0 \\ 0 & a & 0 \\ 0 & 0 & z \end{bmatrix} + \begin{bmatrix} 0 & b & 0 \\ -b & 0 & 0 \\ 0 & 0 & 0 \end{bmatrix}\right) = \exp\left(\begin{bmatrix} a & 0 & 0 \\ 0 & a & 0 \\ 0 & 0 & z \end{bmatrix}\right)\exp\left(\begin{bmatrix} 0 & b & 0 \\ -b & 0 & 0 \\ 0 & 0 & 0 \end{bmatrix}\right)$$

$$= \begin{bmatrix} e^a & 0 & 0 \\ 0 & e^a & 0 \\ 0 & 0 & e^z \end{bmatrix}\begin{bmatrix} \cos b & \sin b & 0 \\ -\sin b & \cos b & 0 \\ 0 & 0 & 1 \end{bmatrix}$$

By closure we have:

---

[54] Section 5 - Chapter 1 - Theorem 5

$$\begin{bmatrix} u & v & 0 \\ -v & u & 0 \\ 0 & 0 & w \end{bmatrix} = \begin{bmatrix} e^a & 0 & 0 \\ 0 & e^a & 0 \\ 0 & 0 & e^z \end{bmatrix} \begin{bmatrix} \cos b & \sin b & 0 \\ -\sin b & \cos b & 0 \\ 0 & 0 & 1 \end{bmatrix}$$

The determinants are:

$$\left(u^2 + v^2\right)w = r^2 e^z$$

$$u^2 + v^2 = r^2$$

Whenever we attempt to calculate a 3-dimensional distance, it collapses by cancellation into a 2-dimensional calculation.

The right hand matrix is the rotation matrix of the seemingly 3-dimensional space that we see about us. It contains upon its leading diagonal the rotation matrix of 2-dimensional euclidean space and the rotation matrix of 1-dimensional space. It is thus 2-dimensional euclidean space together with 1-dimensional space. It is not a bona-fide 3-dimensional space. We note that, when $z = 0$, this is a 2-dimensional space.

The reader might expect that this matrix of everyday 3-dimensional space would give the distance function:

$$d_3^2 = a^2 + b^2 + z^2$$

to which we are accustomed in our everyday lives, but it does not. The understanding that this distance function is the distance function of everyday 3-dimensional space is in error. The error is the undeclared and incorrect assumption that space is fixed and absolute. We set some axes and calculate a 2-dimensional distance:

$$d_2^2 = a^2 + b^2$$

We then repeat the procedure to move into 3-dimensions:

$$d_3^2 = d_2^2 + z^2$$

In both cases, we did a calculation in 2-dimensional space. The illusion that we are calculating a 3-dimensional distance arises from keeping the axes absolutely set. With reference to floors and walls, this works practically, but without the floors and walls to provide an absolute reference frame, we are always calculating in 2-dimensional space.

The conventionally accepted form of 4-dimensional space is:

$$\begin{bmatrix} a & b & 0 & 0 \\ -b & a & 0 & 0 \\ 0 & 0 & c & d \\ 0 & 0 & d & c \end{bmatrix}$$

These matrices are:
Inclusive of the additive identity
Inclusive of additive inverses
Additively closed
Multiplicatively closed
Inclusive of multiplicative inverses

Inclusive of multiplicative identity

Multiplicatively commutative

Non-singular if $c > d$ and $\{a,b\} \neq 0$ unless $\{c,d\} = 0$

With the restrictions, these matrices are an algebraic field.

We split this into two parts:

$$
\begin{bmatrix} a & 0 & 0 & 0 \\ 0 & a & 0 & 0 \\ 0 & 0 & c & 0 \\ 0 & 0 & 0 & c \end{bmatrix}
+
\begin{bmatrix} 0 & b & 0 & 0 \\ -b & 0 & 0 & 0 \\ 0 & 0 & 0 & d \\ 0 & 0 & d & 0 \end{bmatrix}
$$

We note that these two parts are multiplicatively commutative, and so we can exponentiate this matrix and split it.

We have:

$$
\exp\left(
\begin{bmatrix} a & 0 & 0 & 0 \\ 0 & a & 0 & 0 \\ 0 & 0 & c & 0 \\ 0 & 0 & 0 & c \end{bmatrix}
+
\begin{bmatrix} 0 & b & 0 & 0 \\ -b & 0 & 0 & 0 \\ 0 & 0 & 0 & d \\ 0 & 0 & d & 0 \end{bmatrix}
\right)
$$

$$
=
\begin{bmatrix} e^a & 0 & 0 & 0 \\ 0 & e^a & 0 & 0 \\ 0 & 0 & e^c & 0 \\ 0 & 0 & 0 & e^c \end{bmatrix}
\begin{bmatrix} \cos b & \sin b & 0 & 0 \\ -\sin b & \cos b & 0 & 0 \\ 0 & 0 & \cosh d & \sinh d \\ 0 & 0 & \sinh d & \cosh d \end{bmatrix}
$$

By closure we have:

$$
\begin{bmatrix} z & y & 0 & 0 \\ -y & z & 0 & 0 \\ 0 & 0 & x & w \\ 0 & 0 & w & x \end{bmatrix}
=
\begin{bmatrix} e^a & 0 & 0 & 0 \\ 0 & e^a & 0 & 0 \\ 0 & 0 & e^c & 0 \\ 0 & 0 & 0 & e^c \end{bmatrix}
\begin{bmatrix} \cos b & \sin b & 0 & 0 \\ -\sin b & \cos b & 0 & 0 \\ 0 & 0 & \cosh d & \sinh d \\ 0 & 0 & \sinh d & \cosh d \end{bmatrix}
$$

The determinants are:

$$
\left(z^2 + y^2\right)\left(x^2 - w^2\right) = r^2 h^2
$$

The right hand matrix is the rotation matrix of the space that is conventionally called space-time. It contains upon its leading diagonal the rotation matrix of 2-dimensional euclidean space and the rotation matrix of 2-dimensional space-time. It is thus two 2-dimensional spaces stuck together. It is not a bona-fide 4-dimensional space.

Introduction to Section 6

Each of the 1-dimensional and 2-dimensional natural algebras has within it a natural space. The real numbers hold within them the 1-dimensional space; the euclidean complex numbers contain the 2-dimensional euclidean space, and the hyperbolic complex numbers contain the 2-dimensional hyperbolic space. We have found other natural algebras, but do they contain a space, and, if so, how will we recognize it as a space? What mathematical objects need be on our page before we can proclaim that what we have is a natural space?

In this section, we seek to define what we mean by natural space. We do this by listing the mathematical objects, and the properties of those objects, that we feel are sufficient and necessary to make natural space.

When we have the mathematical objects that constitute a natural space, it does not mean the natural space will manifest itself as a geometric space.

Part of this section (addendum 1 to chapter 3) is a lengthy formal definition of what we mean by a trigonometric function. There is much in that definition that we do not meet until later in this work. The definition is included in this section because this is the proper place for it, but the reader is advised to skip reading it in the first reading of this work.

Section 6 : Chapter 1

The Definition of the Trigonometric Functions

In 2-dimensions, there are two types of flat space associated with the natural algebras. There is the 2-dimensional euclidean space associated with the euclidean complex numbers, and there is the 2-dimensional hyperbolic space associated with the hyperbolic complex numbers. Associated with each of these spaces is a set of trigonometric functions. These are the euclidean trigonometric functions, $\{\cos, \sin\}$, together with the various other 'mixtures' of these two functions, and the hyperbolic trigonometric functions, $\{\cosh, \sinh\}$, together with the various other 'mixtures' of these two functions. What is it about these functions that justifies the use of the adjective 'trigonometric'? What properties is a function required to have, or what need it be, before it can be called a trigonometric function?

The 2-dimensional euclidean trigonometric functions, $\{\cos, \sin\}$, were first defined in 6th century India. They were defined to be the projections from a point on the unit circle on to each of the axes of the euclidean plane. This has been the standard definition used by geometers since then. In the case of 2-dimensional euclidean geometry, the *cos* is the projection from the unit circle on to the horizontal axis and the *sin* is the projection from the unit circle on to the vertical axis. In the case of 2-dimensional hyperbolic geometry, the *cosh* is the projection from the hyperbolic unit circle on to the horizontal axis and the *sinh* is the projection from the hyperbolic unit circle on to the vertical axis. Of course, if we draw the hyperbolic unit circle on 2-dimensional euclidean paper, it appears as a hyperbola.

This definition does not fit 3-dimensional space automatically. We can easily extend the unit circle to be a unit sphere (and analogously so for higher dimensional spaces), but a projection parallel to an axis from a point on the 3-dimensional unit sphere might fall on to a plane formed by two other axes and not on to a particular axis. Such a projection is a point. If we want to project from the 3-dimensional unit sphere on to an axis, we need to first project down to a point on a plane and then project down out of that plane to a point on the axis. Combining these two projections together will project down from a point on the unit sphere to a point on an axis. In general, the projection is not parallel to a particular axis. In a 4-dimensional space, we would require to do three axis parallel projections to arrive at an axis.

Nomenclature:
Simple Trigonometric Function
A projection function that is a single projection from the unit sphere is called a simple trigonometric function.

Compound Trigonometric Function
A projection function that is a combination of projections from the unit sphere on to an axis of the space is called a compound trigonometric function.

We proceed with a prospective geometric definition of a trigonometric function.

Prospective Geometric Definition:
Trigonometric Function
A trigonometric function is a projection function from the n-dimensional unit sphere on to an axis – a compound trigonometric function.

Clearly, every 3-dimensional space has three axes and hence will have associated with it a set of three compound trigonometric functions that are projections on to axes. Clearly, every 3-dimensional space also has three planes and hence will have associated with it another set of three simple trigonometric functions that are projections on to planes.

The above definition catches the essential geometric properties of the trigonometric functions $\{\cos, \sin, \cosh, \sinh\}$. However, these functions have other aspects to them. They appear in the rotation matrices of the space with which they are associated. Indeed, the elements of the rotation matrices are all trigonometric functions. We seek to capture this property of the trigonometric functions with a prospective rotation definition.

Prospective Rotation Definition:
Trigonometric Function
A trigonometric function is a function that appears in a rotation matrix.

Of course, if the space does not have a rotation matrix, it cannot have any trigonometric functions. This does not concern us.

In fact, these definitions are not different. If we rotate the position matrix $\begin{bmatrix} 1 & 0 \\ 0 & 1 \end{bmatrix}$, the new position matrix will have, as elements, projections on to the axes from the unit sphere (circle). If the rotation matrix did not contain such projection functions, it could not be a rotation matrix.

The trigonometric functions have another property that we seek to capture in our definition. In 2-dimensions, the rate of change of the projection on to one axis with respect to the change in angle from that axis is the projection on to the other axis. In the case of the 2-dimensional hyperbolic functions, we have:

$$\left\{ \frac{d(\cosh x)}{dx} = \sinh x, \frac{d(\sinh x)}{dx} = \cosh x \right\}$$

with similar, slightly more complicated, relations for the 2-dimensional euclidean functions. In 2-dimensions, the trigonometric functions have series expansions that are splittings of the exponential series (in the euclidean case, with some minus signs chucked in). In the case of the 2-dimensional hyperbolic functions, we have:

$$\cosh x = \frac{x^0}{0!} + \frac{x^2}{2!} + \frac{x^4}{4!} + \dots$$

$$\sinh x = \frac{x^1}{1!} + \frac{x^3}{3!} + \frac{x^5}{5!} + \dots$$

The essence is that a set of trigonometric functions has a differentiation cycle. Thus we have a prospective calculus definition of a trigonometric function.

Prospective Calculus Definition:
Trigonometric Function
A set of trigonometric functions is a set of functions that have a differentiation cycle.

Finally, we decide to cheat. We define a trigonometric function to be that which is an element of the matrices with determinant unity that are produced when a natural algebra is exponentiated. It turns out that this definition satisfies all the previous prospective definitions.

Cheating Definition:
Trigonometric Function
A trigonometric function is an element of the matrices with determinant unity that are produced when a natural algebra is exponentiated.

However, hindsight allows us to give a precise form of this definition. This precise definition is given as an addendum to this chapter, and this precise definition is the definition that we take to be the proper definition of a trigonometric function. This precise definition satisfies all the prospective definitions given above. The functions it defines are all projection functions that appear in rotation matrices, that have differentiation cycles, and that drop out of the natural algebras.

Section 6 : Addendum 1 to Chapter 1
Part 1
Definition of Simple Trigonometric Functions[55]

We first deal with the functions that are elements of the rotation matrices that occur in the simple polar form of algebraic matrix forms. These are the simple trigonometric functions.

The basic trigonometric functions of 2-dimensional space have series expansions that are 2-way splittings of the exponential series. These splittings of the exponential series have hypergeometric representation.

$$\cos x = hypergeom\left([\ ],\left[\frac{1}{2}\right],-\left(\frac{x}{2}\right)^2\right) = \frac{x^0}{0!} - \frac{x^2}{2!} + \frac{x^4}{4!} - \frac{x^6}{6!} + \dots$$

$$\sin x = x\left(hypergeom\left([\ ],\left[\frac{3}{2}\right],-\left(\frac{x}{2}\right)^2\right)\right) = \frac{x^1}{1!} - \frac{x^3}{3!} + \frac{x^5}{5!} - \frac{x^7}{7!} + \dots$$

$$\cosh x = hypergeom\left([\ ],\left[\frac{1}{2}\right],\left(\frac{x}{2}\right)^2\right) = \frac{x^0}{0!} + \frac{x^2}{2!} + \frac{x^4}{4!} + \frac{x^6}{6!} + \dots$$

$$\sinh x = x\left(hypergeom\left([\ ],\left[\frac{3}{2}\right],\left(\frac{x}{2}\right)^2\right)\right) = \frac{x^1}{1!} + \frac{x^3}{3!} + \frac{x^5}{5!} + \frac{x^7}{7!} + \dots$$

The elements in the rotation matrices of the simple polar form of the 3-dimensional algebraic matrix forms are 3-way splittings of the exponential series with hypergeometric representation:

$$AH_3(x) = hypergeom\left([\ ],\left[\frac{1}{3},\frac{2}{3}\right],\left(\frac{x}{3}\right)^3\right) = \frac{x^0}{0!} + \frac{x^3}{3!} + \frac{x^6}{6!} + \frac{x^9}{9!} + \dots$$

$$BH_3(x) = \frac{x}{1!}\left(hypergeom\left([\ ],\left[\frac{2}{3},\frac{4}{3}\right],\left(\frac{x}{3}\right)^3\right)\right) = \frac{x^1}{1!} + \frac{x^4}{4!} + \frac{x^7}{7!} + \frac{x^{10}}{10!} + \dots$$

$$CH_3(x) = \frac{x^2}{2!}\left(hypergeom\left([\ ],\left[\frac{4}{3},\frac{5}{3}\right],\left(\frac{x}{3}\right)^3\right)\right) = \frac{x^2}{2!} + \frac{x^5}{5!} + \frac{x^8}{8!} + \frac{x^{11}}{11!} + \dots$$

---

[55]At a first reading, much of this material will be meaningless to the reader. We are dealing with things that are not revealed until later in this work. The reader is urged to skip through these definitions and proceed to the next chapter.

$$AE_3(x) = hypergeom\left(\left[\ \right],\left[\frac{1}{3},\frac{2}{3}\right],-\left(\frac{x}{3}\right)^3\right) = \frac{x^0}{0!} - \frac{x^3}{3!} + \frac{x^6}{6!} - \frac{x^9}{9!} + \dots$$

$$BE_3(x) = \frac{x}{1!}\left(hypergeom\left(\left[\ \right],\left[\frac{2}{3},\frac{4}{3}\right],-\left(\frac{x}{3}\right)^3\right)\right) = \frac{x^1}{1!} - \frac{x^4}{4!} + \frac{x^7}{7!} - \frac{x^{10}}{10!} + \dots$$

$$CE_3(x) = \frac{x^2}{2!}\left(hypergeom\left(\left[\ \right],\left[\frac{4}{3},\frac{5}{3}\right],-\left(\frac{x}{3}\right)^3\right)\right) = \frac{x^2}{2!} - \frac{x^5}{5!} + \frac{x^8}{8!} - \frac{x^{11}}{11!} + \dots$$

The elements in the rotation matrices of the simple polar form of the 4-dimensional algebraic matrix forms are 4-way splittings of the exponential series with hypergeometric representation:

$$AH_4(x) = hypergeom\left(\left[\ \right],\left[\frac{1}{4},\frac{1}{2},\frac{3}{4}\right],\left(\frac{x}{4}\right)^4\right) = 1 + \frac{x^4}{4!} + \frac{x^8}{8!} + \frac{x^{12}}{12!} + \dots$$

$$BH_4(x) = x\left(hypergeom\left(\left[\ \right],\left[\frac{1}{2},\frac{3}{4},\frac{5}{4}\right],\left(\frac{x}{4}\right)^4\right)\right) = x + \frac{x^5}{5!} + \frac{x^9}{9!} + \frac{x^{13}}{13!} + \dots$$

$$CH_4(x) = \frac{x^2}{2}\left(hypergeom\left(\left[\ \right],\left[\frac{3}{4},\frac{5}{4},\frac{6}{4}\right],\left(\frac{x}{4}\right)^4\right)\right) = \frac{x^2}{2!} + \frac{x^6}{6!} + \frac{x^{10}}{10!} + \dots$$

$$DH_4(x) = \frac{x^3}{6}\left(hypergeom\left(\left[\ \right],\left[\frac{5}{4},\frac{6}{4},\frac{7}{4}\right],\left(\frac{x}{4}\right)^4\right)\right) = \frac{x^3}{3!} + \frac{x^7}{7!} + \frac{x^{11}}{11!} + \dots$$

$$AE_4(x) = hypergeom\left(\left[\ \right],\left[\frac{1}{4},\frac{1}{2},\frac{3}{4}\right],-\left(\frac{x}{4}\right)^4\right) = 1 - \frac{x^4}{4!} + \frac{x^8}{8!} - \frac{x^{12}}{12!} + \dots$$

$$BE_4(x) = x\left(hypergeom\left(\left[\ \right],\left[\frac{1}{2},\frac{3}{4},\frac{5}{4}\right],-\left(\frac{x}{4}\right)^4\right)\right) = x - \frac{x^5}{5!} + \frac{x^9}{9!} - \frac{x^{13}}{13!} + \dots$$

$$CE_4(x) = \frac{x^2}{2}\left(hypergeom\left(\left[\ \right],\left[\frac{3}{4},\frac{5}{4},\frac{6}{4}\right],-\left(\frac{x}{4}\right)^4\right)\right) = \frac{x^2}{2!} - \frac{x^6}{6!} + \frac{x^{10}}{10!} - \dots$$

$$DE_4(x) = \frac{x^3}{6}\left(hypergeom\left(\left[\ \right],\left[\frac{5}{4},\frac{6}{4},\frac{7}{4}\right],-\left(\frac{x}{4}\right)^4\right)\right) = \frac{x^3}{3!} - \frac{x^7}{7!} + \frac{x^{11}}{11!} - \dots$$

The pattern within the hypergeometric format is clear. By similar means, we can define the simple trigonometric function for *n*-dimensions as *n*-ways splittings of the exponential.

Definition:
Simple Trigonometric Function

The hypergeometric representation of the H-type functions is always of the form:

$$AH_n = \frac{x^0}{0!}\left(hypergeom\left([\ ], \left[\frac{1}{n}, \frac{2}{n}, ... \frac{n-1}{n}\right], \left(\frac{x}{n}\right)^n\right)\right) = \frac{x^0}{0!} + \frac{x^n}{n!} + ...$$

$$BH_n = \frac{x^1}{1!}\left(hypergeom\left([\ ], \left[\frac{2}{n}, \frac{3}{n}, ..., \frac{n-1}{n}, \frac{n+1}{n}\right], \left(\frac{x}{n}\right)^n\right)\right) = \frac{x^1}{1!} + \frac{x^{n+1}}{(n+1)!} + ...$$

...

$$\Theta H_n = \frac{x^{n-1}}{(n-1)!}\left(hypergeom\left([\ ], \left[\frac{n+1}{n}, \frac{n+2}{n}, ..., \frac{2n-1}{n}\right], \left(\frac{x}{n}\right)^n\right)\right) = \frac{x^{n-1}}{(n-1)!} + \frac{x^{2n-1}}{(2n-1)!} +$$

where ΘH_n is the *n*-ways splitting of the exponential series whose first term is x^{n-1}.
Note that the term $\frac{n}{n}$ is always absent from the middle bracket.

The hypergeometric representation of the E-type functions is always of the form:

$$AE_n = \frac{x^0}{0!}\left(hypergeom\left([\ ], \left[\frac{1}{n}, \frac{2}{n}, ... \frac{n-1}{n}\right], -\left(\frac{x}{n}\right)^n\right)\right) = \frac{x^0}{0!} - \frac{x^n}{n!} + ...$$

$$BE_n = \frac{x^1}{1!}\left(hypergeom\left([\ ], \left[\frac{2}{n}, \frac{3}{n}, ..., \frac{n-1}{n}, \frac{n+1}{n}\right], -\left(\frac{x}{n}\right)^n\right)\right) = \frac{x^1}{1!} - \frac{x^{n+1}}{(n+1)!} + ...$$

...

$$\Theta E_n = \frac{x^{n-1}}{(n-1)!}\left(hypergeom\left([\ ], \left[\frac{n+1}{n}, \frac{n+2}{n}, ..., \frac{2n-1}{n}\right], -\left(\frac{x}{n}\right)^n\right)\right) = \frac{x^{n-1}}{(n-1)!} - \frac{x^{2n-1}}{(2n-1)!} +$$

Of course:

$$hypergeom\left([\ ],[\ ], \left(\frac{x}{1}\right)^1\right) = e^x \qquad \& \qquad hypergeom\left([\ ],[\ ], -\left(\frac{x}{1}\right)^1\right) = e^{-x}$$

We do not need the hypergeometric form of these functions; it is included to give precision to our definition and to emphasize the similar nature of these functions to the 2-dimensional trigonometric functions.

Notes:

We always have:

$$AE_n = AH_{2n} - \Theta H_{2n}$$

where ΘH_{2n} is the $2n$-ways splitting of the exponential series whose first term is x^{2n-1}, with similar relations for $BE_n, CE_n, \ldots$ We always have:

$$AH_n = AH_{2n} + \Theta H_{2n}$$

where ΘH_{2n} is the $2n$-ways splitting of the exponential series whose first term is x^{2n-1}, with similar relations for $BH_n, CH_n, \ldots$ Similar relations can be derived (think of the series expansions) from the addition of more than two simple trigonometric functions, for example:

$$AH_3 = AH_9 + DH_9 + GH_9$$

So it is that the higher dimensional simple trigonometric functions are 'folded' inside the lower dimensional simple trigonometric functions. We always have:

$$XH_{2n} = \frac{1}{2}\left(XH_n + XE_n\right)$$

where X represents any of the simple trigonometric functions. So it is that, once we have established both the H-type and the E-type simple trigonometric functions for a particular n, the higher dimensional H-type simple trigonometric functions $2n$, $4n$, $8n$ ... are determined. Of course, the H-type simple trigonometric functions of dimension n where n is odd are not so determined. Re-arranging this equation gives:

$$XE_n = 2XH_{2n} - XH_n$$

So it is that once we have established the H-type simple trigonometric functions for a particular n and $2n$, the n-dimensional E-type functions are determined.

Part 2

Definition of Compound Trigonometric Functions

(We will also use the term v-functions to mean compound trigonometric functions.)

We now deal with the functions that are the elements of the single compound rotation matrix that occurs in the compound polar form of algebraic matrix forms. These are the compound trigonometric functions.

The compound trigonometric functions occur in the rotation matrix that is the product of the simple rotation matrices in the simple polar form of an algebraic matrix form. An example: The simple polar form of the 3-dimensional algebraic matrix form, $C_3 L^1 H^2{}_{(j=1,k=1)}$, is:

$$\exp\left(\begin{bmatrix} a & b & c \\ c & a & b \\ b & c & a \end{bmatrix}\right) =$$

$$\begin{bmatrix} r & 0 & 0 \\ 0 & r & 0 \\ 0 & 0 & r \end{bmatrix} \begin{bmatrix} AH_3(b) & BH_3(b) & CH_3(b) \\ CH_3(b) & AH_3(b) & BH_3(b) \\ BH_3(b) & CH_3(b) & AH_3(b) \end{bmatrix} \begin{bmatrix} AH_3(c) & CH_3(c) & BH_3(c) \\ BH_3(c) & AH_3(c) & CH_3(c) \\ CH_3(c) & BH_3(c) & AH_3(c) \end{bmatrix}$$

We set:

$$v_{\left[C_3 L^1 H^2{}_{(j=1,k=1)}\right]} A(b,c) = AH_3(b)AH_3(c) + BH_3(b)BH_3(c) + CH_3(b)CH_3(c)$$

$$v_{\left[C_3 L^1 H^2{}_{(j=1,k=1)}\right]} B(b,c) = AH_3(c)BH_3(b) + AH_3(b)CH_3(c) + BH_3(c)CH_3(b)$$

$$v_{\left[C_3 L^1 H^2{}_{(j=1,k=1)}\right]} C(b,c) = AH_3(c)CH_3(b) + AH_3(b)BH_3(c) + BH_3(b)CH_3(c)$$

Leading to the compound polar form:

$$\exp\left(\begin{bmatrix} a & b & c \\ c & a & b \\ b & c & a \end{bmatrix}\right) =$$

$$\begin{bmatrix} r & 0 & 0 \\ 0 & r & 0 \\ 0 & 0 & r \end{bmatrix} \begin{bmatrix} v_{\left[C_3 L^1 H^2{}_{(j=1,k=1)}\right]} A(b,c) & v_{\left[C_3 L^1 H^2{}_{(j=1,k=1)}\right]} B(b,c) & v_{\left[C_3 L^1 H^2{}_{(j=1,k=1)}\right]} C(b,c) \\ v_{\left[C_3 L^1 H^2{}_{(j=1,k=1)}\right]} C(b,c) & v_{\left[C_3 L^1 H^2{}_{(j=1,k=1)}\right]} A(b,c) & v_{\left[C_3 L^1 H^2{}_{(j=1,k=1)}\right]} B(b,c) \\ v_{\left[C_3 L^1 H^2{}_{(j=1,k=1)}\right]} B(b,c) & v_{\left[C_3 L^1 H^2{}_{(j=1,k=1)}\right]} C(b,c) & v_{\left[C_3 L^1 H^2{}_{(j=1,k=1)}\right]} A(b,c) \end{bmatrix}$$

The functions in the rotation matrix are the compound trigonometric functions of this particular algebra. We call them the v-functions.

Section 6 : Chapter 2

The Definition of Rotation

In 2-dimensions, we have two types of rotation, euclidean rotation and hyperbolic rotation. Euclidean rotation is a multi-valued thing in that $\{0, 2\pi, 4\pi, ...\}$ are the same angle. Nothing like this occurs in hyperbolic rotation. We seek to define 'rotation' and 'rotation matrix' in such a way that includes everything that these two rotations have in common but excludes everything that they do not have in common.

A rotation is a movement in space. As such it is a linear transformation. As such it can be represented as a matrix. The properties of a rotation are:

i) The distance of a position (as calculated using the appropriate distance function) from the origin is the same after the rotation as it was before the rotation. Technically speaking, we say that the rotational linear transformation leaves the distance invariant.

ii) Two successive rotations through angles α and β are the same as a single rotation through the angle $(\alpha + \beta)$.

iii) The elements within a rotation matrix are trigonometric functions.

In general, the distance functions of the n-dimensional natural algebraic spaces are the nth roots[56] of the determinants of the matrix representations of the associated n-dimensional natural algebras. The requirement that a rotation matrix leaves the metric invariant is thus the requirement that the rotation matrix has determinant equal to unity[57].

$Rot(\alpha + \beta) = Rot(\alpha) Rot\beta$ means that rotation is of an exponential nature - $p^{a+b} = p^a p^b$, and we will have to use exponentiation to construct the rotation matrices.

With this understanding, we write our definition of rotation.

Definition:

Rotation

A rotation matrix, A, is such that:

i) Multiplication by it leaves the metric invariant. That is: $\det(A) = 1$.

ii) $A(\alpha) \bullet A(\beta) = A(\alpha + \beta)$

iii) The elements of A are all trigonometric functions.

The requirement that the elements within a rotation matrix are trigonometric functions and the requirement that the determinant of a rotation matrix be unity will automatically lead to trigonometric identities similar to:

---

[56] We have not yet approached this aspect of the algebras in more than two dimensions.
[57] We require that the determinant of a position matrix is the same after multiplication by the rotation matrix as it was prior to that multiplication.

$$\det\left(\begin{bmatrix} \cos\alpha & \sin\alpha \\ -\sin\alpha & \cos\alpha \end{bmatrix}\right) = \cos^2\alpha + \sin^2\alpha = 1$$

Constructing the rotation matrices by exponentiation will automatically lead to multiple angle identities similar to:

$$\exp\left(\begin{bmatrix} 0 & b+c \\ -(b+c) & 0 \end{bmatrix}\right) = \begin{bmatrix} \cos(b+c) & \sin(b+c) \\ -\sin(b+c) & \cos(b+c) \end{bmatrix}$$

$$\exp\left(\begin{bmatrix} 0 & b+c \\ -(b+c) & 0 \end{bmatrix}\right) = \exp\left(\begin{bmatrix} 0 & b \\ -b & 0 \end{bmatrix}\right)\exp\left(\begin{bmatrix} 0 & c \\ -c & 0 \end{bmatrix}\right)$$

$$= \begin{bmatrix} \cos b & \sin b \\ -\sin b & \cos b \end{bmatrix}\begin{bmatrix} \cos c & \sin c \\ -\sin c & \cos c \end{bmatrix}$$

$$= \begin{bmatrix} \cos b\cos c - \sin b\sin c & \cos b\sin c + \sin b\cos c \\ \sim & \sim \end{bmatrix}$$

$$\cos(b+c) = \cos b\cos c - \sin b\sin c$$

$$\sin(b+c) = \cos b\sin c + \sin b\cos c$$

Section 6 : Chapter 3

The Definition of the Distance Function

We would like our distance functions to be called metrics, but that word is already in use with a precise meaning that is different from what we mean by distance function. Although some of the distance functions we will meet satisfy the axioms of a metric, not all of them will. Since the 2-dimensional hyperbolic space-time of special relativity, and hence of reality, does not satisfy all the metric axioms, this is not necessarily a great loss.

Definition:
Distance Function
The distance function of a natural space is the nth root of the determinant of the associated algebraic matrix form.

This definition is forced upon us by the polar forms of the algebras. It is quite different from the spaces to which we are accustomed.

The determinant of a matrix is invariant under change of basis. We view a $n \times n$ algebraic matrix form as a position vector in n-dimensional natural space. The length of this vector is the only property that is invariant under change of basis.

Section 6 : Chapter 4

The Definition of Natural Space

We define natural space to be a particular bundle of mathematical objects. This bundle includes a distance function, a set of rotation matrices, and a set of trigonometric functions. Nothing less than this will suffice, and we contend that nothing more is required to make a natural space.

We require the trigonometric functions to be the ones that appear in the rotation matrices. We require the distance function to have as many variables as there are different trigonometric functions within the set of compound (or simple) trigonometric functions; this means as many axes as there are projections on to those axes. We require that the distance function declare itself to be a distance function rather than we having to impose this nature upon it.

Definition:

n-dimensional Natural Space

i) A Distance Function

ii) A set of $(n-1)$ Simple Rotation Matrices

iii) A set of n Simple Trigonometric Functions

Note: The set of simple rotation matrices will generate a compound rotation matrix.
Note: The set of simple trigonometric functions will generate a set of compound trigonometric functions.

The above mathematical objects are exactly what fall out of the natural algebras.

Introduction to Section 7

In this section, we meet the 3-dimensional natural algebras. With them, we meet the 3-dimensional natural spaces together with the 3-dimensional trigonometric functions. These 3-dimensional spaces are nothing like the 3-dimensional space to which we are accustomed. The 3-dimensional trigonometric functions are likely to be new to the first time reader who might have thought that there were only four basic trigonometric functions $\{\sin, \cos, \sinh, \cosh\}$. We also meet the concepts of shadow algebras and shadow spaces.

Section 7 : Chapter 1

The Polar Form of the $C_3 L^1 H^2_{(j=1,k=1)}$ Algebra

We have previously established that the only forms of 3×3 matrices with three independent variables that are algebraic fields are (with restrictions):

$$C_3 = \begin{bmatrix} a & b & c \\ jc & a & kb \\ jb & \left(\dfrac{j}{k}\right)c & a \end{bmatrix} : j \in \mathbb{R}, k \in \mathbb{R} \neq 0$$

With $\{j,k\} = \pm 1$, the only flat forms of 3×3 matrices (with three independent variables) that are an algebraic field are (with restrictions):

$$C_3 L^1 H^2_{(j=1,k=1)} = \begin{bmatrix} a & b & c \\ c & a & b \\ b & c & a \end{bmatrix}, \qquad C_3 L^1 E^1 H^1_{(j=-1,k=1)} = \begin{bmatrix} a & b & c \\ -c & a & b \\ -b & -c & a \end{bmatrix}$$

$$C_3 L^1 E^2_{(j=1,k=-1)} = \begin{bmatrix} a & b & c \\ c & a & -b \\ b & -c & a \end{bmatrix}, \quad C_3 L^1 E^1 H^1_{(j=-1,k=-1)} = \begin{bmatrix} a & b & c \\ -c & a & -b \\ -b & c & a \end{bmatrix}$$

We have:

$$\exp\left(\begin{bmatrix} a & b & c \\ c & a & b \\ b & c & a \end{bmatrix}\right) = \exp\left(\begin{bmatrix} a & 0 & 0 \\ 0 & a & 0 \\ 0 & 0 & a \end{bmatrix} + \begin{bmatrix} 0 & b & 0 \\ 0 & 0 & b \\ b & 0 & 0 \end{bmatrix} + \begin{bmatrix} 0 & 0 & c \\ c & 0 & 0 \\ 0 & c & 0 \end{bmatrix}\right)$$

$$= \exp\left(\begin{bmatrix} a & 0 & 0 \\ 0 & a & 0 \\ 0 & 0 & a \end{bmatrix}\right) \exp\left(\begin{bmatrix} 0 & b & 0 \\ 0 & 0 & b \\ b & 0 & 0 \end{bmatrix}\right) \exp\left(\begin{bmatrix} 0 & 0 & c \\ c & 0 & 0 \\ 0 & c & 0 \end{bmatrix}\right)$$

$$= PROD \begin{pmatrix} \begin{bmatrix} e^a & 0 & 0 \\ 0 & e^a & 0 \\ 0 & 0 & e^a \end{bmatrix} \\ \begin{bmatrix} 1+\dfrac{b^3}{3!}+\dfrac{b^6}{6!}+... & b+\dfrac{b^4}{4!}+\dfrac{b^7}{7!}+... & \dfrac{b^2}{2!}+\dfrac{b^5}{5!}+... \\ \dfrac{b^2}{2!}+\dfrac{b^5}{5!}+... & 1+\dfrac{b^3}{3!}+\dfrac{b^6}{6!}+... & b+\dfrac{b^4}{4!}+\dfrac{b^7}{7!}+... \\ b+\dfrac{b^4}{4!}+\dfrac{b^7}{7!}+... & \dfrac{b^2}{2!}+\dfrac{b^5}{5!}+... & 1+\dfrac{b^3}{3!}+\dfrac{b^6}{6!}+... \end{bmatrix} \\ \begin{bmatrix} 1+\dfrac{c^3}{3!}+\dfrac{c^6}{6!}+... & \dfrac{c^2}{2!}+\dfrac{c^5}{5!}+... & c+\dfrac{c^4}{4!}+\dfrac{c^7}{7!}+... \\ c+\dfrac{c^4}{4!}+\dfrac{c^7}{7!}+... & 1+\dfrac{c^3}{3!}+\dfrac{c^6}{6!}+... & \dfrac{c^2}{2!}+\dfrac{c^5}{5!}+... \\ \dfrac{c^2}{2!}+\dfrac{c^5}{5!}+... & c+\dfrac{c^4}{4!}+\dfrac{c^7}{7!}+... & 1+\dfrac{c^3}{3!}+\dfrac{c^6}{6!}+... \end{bmatrix} \end{pmatrix}$$

The infinite series within the second two matrices are 3-way splittings of the exponential series. Thus, by the strict definition of a trigonometric function (the hyper-geometric one), these are the simple trigonometric functions of this algebra. We call these functions the 3-dimensional H-type[58] simple trigonometric functions, and we denote them:

$$AH_3(x) = 1+\frac{x^3}{3!}+\frac{x^6}{6!}+...$$

$$BH_3(x) = x+\frac{x^4}{4!}+\frac{x^7}{7!}+...$$

$$CH_3(x) = \frac{x^2}{2!}+\frac{x^5}{5!}+\frac{x^8}{8!}+...$$

Leading to:

$$\exp\left(\begin{bmatrix} a & b & c \\ c & a & b \\ b & c & a \end{bmatrix}\right) =$$

---

[58] The H is taken from hyperbolic with an eye on the 2-dimensional hyperbolic trigonometric functions. We take the view that hyperbolic space is a 2-dimensional thing and that we cannot apply that adjective to any spaces other than the 2-dimensional one – hence the H.

$$\begin{bmatrix} r & 0 & 0 \\ 0 & r & 0 \\ 0 & 0 & r \end{bmatrix} \begin{bmatrix} AH_3(b) & BH_3(b) & CH_3(b) \\ CH_3(b) & AH_3(b) & BH_3(b) \\ BH_3(b) & CH_3(b) & AH_3(b) \end{bmatrix} \begin{bmatrix} AH_3(c) & CH_3(c) & BH_3(c) \\ BH_3(c) & AH_3(c) & CH_3(c) \\ CH_3(c) & BH_3(c) & AH_3(c) \end{bmatrix}$$

The matrices containing the H-type simple trigonometric functions were formed by exponentiating a matrix with $Trace = 0$. Since $\det\left(\exp\left([\ ]\right)\right) = \exp\left(Trace\left([\ ]\right)\right)$, the determinants of these matrices are unity[59]. This leads, in both cases, to the same trigonometric identity analogous to the well known $\cosh^2 x - \sinh^2 x = 1$. In the 3-dimensional case, this identity is:

Trigonometric Identity 1:
$$AH_3{}^3(x) + BH_3{}^3(x) + CH_3{}^3(x) - 3AH_3(x)BH_3(x)CH_3(x) = 1$$

We have:

$$\begin{bmatrix} AH_3(b+x) & BH_3(b+x) & CH_3(b+x) \\ CH_3(b+x) & AH_3(b+x) & BH_3(b+x) \\ BH_3(b+x) & CH_3(b+x) & AH_3(b+x) \end{bmatrix} = \exp\left(\begin{bmatrix} 0 & b+x & 0 \\ 0 & 0 & b+x \\ b+x & 0 & 0 \end{bmatrix}\right)$$

$$= \exp\left(\begin{bmatrix} 0 & b & 0 \\ 0 & 0 & b \\ b & 0 & 0 \end{bmatrix} + \begin{bmatrix} 0 & x & 0 \\ 0 & 0 & x \\ x & 0 & 0 \end{bmatrix}\right)$$

$$= \exp\left(\begin{bmatrix} 0 & b & 0 \\ 0 & 0 & b \\ b & 0 & 0 \end{bmatrix}\right)\exp\left(\begin{bmatrix} 0 & x & 0 \\ 0 & 0 & x \\ x & 0 & 0 \end{bmatrix}\right)$$

$$= PROD\left(\begin{bmatrix} AH_3(b) & BH_3(b) & CH_3(b) \\ CH_3(b) & AH_3(b) & BH_3(b) \\ BH_3(b) & CH_3(b) & AH_3(b) \end{bmatrix} \begin{bmatrix} AH_3(x) & BH_3(x) & CH_3(x) \\ CH_3(x) & AH_3(x) & BH_3(x) \\ BH_3(x) & CH_3(x) & AH_3(x) \end{bmatrix}\right)$$

Thus, this matrix is a rotation matrix. By a similar procedure, the other matrix containing the H-type simple trigonometric functions is also a rotation matrix.

The reader has seen this procedure in the case of a 2-dimensional matrix in the chapter defining a rotation matrix. Clearly, it will work in all cases where a matrix is

---

[59] We will shortly give the sums of these series. With those sums, the fact that the determinant is unity can be shown by direct calculation.

formed by exponentiation. In all cases, it will lead to a trigonometric identity, or three. In the cases of the two matrices considered here, we have:

Trigonometric Identity 2, 3, & 4:

$$AH_3(b+x) = AH_3(b)AH_3(x) + BH_3(b)CH_3(x) + CH_3(b)BH_3(x)$$

$$BH_3(b+x) = AH_3(b)BH_3(x) + BH_3(b)AH_3(x) + CH_3(b)CH_3(x)$$

$$CH_3(b+x) = AH_3(b)CH_3(x) + BH_3(b)BH_3(x) + CH_3(b)AH_3(x)$$

The exponentiated matrix is part of the algebra, and so we have:

$$
\begin{bmatrix} x & y & z \\ z & x & y \\ y & z & x \end{bmatrix} =
$$

$$
\begin{bmatrix} r & 0 & 0 \\ 0 & r & 0 \\ 0 & 0 & r \end{bmatrix}
\begin{bmatrix} AH_3(b) & BH_3(b) & CH_3(b) \\ CH_3(b) & AH_3(b) & BH_3(b) \\ BH_3(b) & CH_3(b) & AH_3(b) \end{bmatrix}
\begin{bmatrix} AH_3(c) & CH_3(c) & BH_3(c) \\ BH_3(c) & AH_3(c) & CH_3(c) \\ CH_3(c) & BH_3(c) & AH_3(c) \end{bmatrix}
$$

$$
\det\left(\begin{bmatrix} x & y & z \\ z & x & y \\ y & z & x \end{bmatrix} \right) =
$$

$$
\det\left(\begin{bmatrix} r & 0 & 0 \\ 0 & r & 0 \\ 0 & 0 & r \end{bmatrix}
\begin{bmatrix} AH_3(b) & BH_3(b) & CH_3(b) \\ CH_3(b) & AH_3(b) & BH_3(b) \\ BH_3(b) & CH_3(b) & AH_3(b) \end{bmatrix}
\begin{bmatrix} AH_3(c) & CH_3(c) & BH_3(c) \\ BH_3(c) & AH_3(c) & CH_3(c) \\ CH_3(c) & BH_3(c) & AH_3(c) \end{bmatrix} \right)
$$

$$x^3 + y^3 + z^3 - 3xyz = r^3.1.1$$

Leading to the distance function:

$$r = \sqrt[3]{x^3 + y^3 + z^3 - 3xyz}$$

This is the distance function forced upon us by the algebra. By the procedure above, it declares itself to be a distance function. This is a function of three independent real variables.

Note:

Since $e^a > 0$ for all values of a, then $r^3 > 0$, and so $x^3 + y^3 + z^3 - 3xyz > 0$.

Hence, non-singularity is guaranteed.

The reader might expect that taking the cube root of a real number leads to one real root and two complex roots. However, this is not the 2-dimensional complex number algebra.

     The H-type simple trigonometric functions appear in the rotation matrix. Consideration of the series representations of the H-type simple trigonometric functions shows that they have a differentiation cycle (of length three).

We have a distance function, a set of two simple rotation matrices, and a set of three simple trigonometric functions. Multiplication of the two simple rotation matrices will give the compound trigonometric functions in the compound rotation matrix. We have a space.

If the series of the simple trigonometric functions are summed using techniques developed within the real number algebra, they have have representation in standard function form:

$$AH_3(x) = \frac{1}{3}\left(e^x + 2e^{-\left(\frac{x}{2}\right)} \cos\left(\frac{\sqrt{3}}{2}x \right) \right)$$

$$BH_3(x) = \frac{1}{3}\left(e^x + e^{-\left(\frac{x}{2}\right)} \left(\sqrt{3}\sin\left(\frac{\sqrt{3}}{2}x \right) - \cos\left(\frac{\sqrt{3}}{2}x \right) \right) \right)$$

$$CH_3(x) = \frac{1}{3}\left(e^x - e^{-\left(\frac{x}{2}\right)} \left(\sqrt{3}\sin\left(\frac{\sqrt{3}}{2}x \right) + \cos\left(\frac{\sqrt{3}}{2}x \right) \right) \right)$$

The reader might be concerned that, in calculating these summations, we are mixing algebras together. Clearly, if the variable were a euclidean complex number, the above summations would be nonsense. However, these are relations within the real number algebra, which is a sub-algebra of this algebra; as such, they are part of the 3-dimensional algebra.

Section 7 : Chapter 2

The Polar Form of the $C_3 L^j E^2_{(j=1,k=-1)}$ Algebra

The polar form of the $C_3 L^j E^2_{(j=1,k=-1)}$ algebra is:

$$\exp\left(\begin{bmatrix} a & b & c \\ c & a & -b \\ b & -c & a \end{bmatrix}\right) =$$

$$PROD \left(\begin{bmatrix} e^a & 0 & 0 \\ 0 & e^a & 0 \\ 0 & 0 & e^a \end{bmatrix} \begin{bmatrix} 1 - \dfrac{b^3}{3!} + \dfrac{b^6}{6!} - \dots & \dfrac{b}{1!} - \dfrac{b^4}{4!} + \dfrac{b^7}{7!} - \dots & -\left(\dfrac{b^2}{2!} - \dfrac{b^5}{5!} + \dfrac{b^8}{8!} - \dots\right) \\ -\left(\dfrac{b^2}{2!} - \dfrac{b^5}{5!} + \dfrac{b^8}{8!} - \dots\right) & 1 - \dfrac{b^3}{3!} + \dfrac{b^6}{6!} - \dots & -\left(\dfrac{b}{1!} - \dfrac{b^4}{4!} + \dfrac{b^7}{7!} - \dots\right) \\ \dfrac{b}{1!} - \dfrac{b^4}{4!} + \dfrac{b^7}{7!} - \dots & \dfrac{b^2}{2!} - \dfrac{b^5}{5!} + \dfrac{b^8}{8!} - \dots & 1 - \dfrac{b^3}{3!} + \dfrac{b^6}{6!} - \dots \end{bmatrix} \begin{bmatrix} 1 - \dfrac{c^3}{3!} + \dfrac{c^6}{6!} - \dots & -\left(\dfrac{c^2}{2!} - \dfrac{c^5}{5!} + \dfrac{c^8}{8!} - \dots\right) & \dfrac{c}{1!} - \dfrac{c^4}{4!} + \dfrac{c^7}{7!} - \dots \\ \dfrac{c}{1!} - \dfrac{c^4}{4!} + \dfrac{c^7}{7!} - \dots & 1 - \dfrac{c^3}{3!} + \dfrac{c^6}{6!} - \dots & \dfrac{c^2}{2!} - \dfrac{c^5}{5!} + \dfrac{c^8}{8!} - \dots \\ -\left(\dfrac{c^2}{2!} - \dfrac{c^5}{5!} + \dfrac{c^8}{8!} - \dots\right) & -\left(\dfrac{c}{1!} - \dfrac{c^4}{4!} + \dfrac{c^7}{7!} - \dots\right) & 1 - \dfrac{c^3}{3!} + \dfrac{c^6}{6!} - \dots \end{bmatrix} \right)$$

The infinite series within the second two matrices are 3-way splittings of the exponential series with some minus signs chucked in. Thus, these are simple trigonometric functions. We call these functions the 3-dimensional [60]E-type simple trigonometric functions, and we denote them:

---

[60] The E is taken from euclidean with an eye on the 2-dimensional euclidean trigonometric functions. We take the view that euclidean space is a 2-dimensional thing and that we cannot apply that adjective to any spaces other than the 2-dimensional one – hence the E.

$$AE_3(x) = 1 - \frac{x^3}{3!} + \frac{x^6}{6!} - \ldots$$

$$BE_3(x) = x - \frac{x^4}{4!} + \frac{x^7}{7!} - \ldots$$

$$CE_3(x) = \frac{x^2}{2!} - \frac{x^5}{5!} + \frac{x^8}{8!} - \ldots$$

Leading to:

$$\exp\left(\begin{bmatrix} a & b & c \\ c & a & -b \\ b & -c & a \end{bmatrix}\right) =$$

$$\begin{bmatrix} r & 0 & 0 \\ 0 & r & 0 \\ 0 & 0 & r \end{bmatrix} \begin{bmatrix} AE_3(b) & BE_3(b) & -CE_3(b) \\ -CE_3(b) & AE_3(b) & -BE_3(b) \\ BE_3(b) & CE_3(b) & AE_3(b) \end{bmatrix} \begin{bmatrix} AE_3(c) & -CE_3(c) & BE_3(c) \\ BE_3(c) & AE_3(c) & CE_3(c) \\ -CE_3(c) & -BE_3(c) & AE_3(c) \end{bmatrix}$$

Again, the matrices containing the E-type simple trigonometric functions were formed by exponentiating a matrix with $Trace = 0$, and the determinants of these matrices are unity. Both matrices lead to the following trigonometric identity.

Trigonometric Identity 1:
$$AE_3{}^3(x) - BE_3{}^3(x) + CE_3{}^3(x) + 3AE_3(x)BE_3(x)CE_3(x) = 1$$

We have:

$$\begin{bmatrix} AE_3(b+x) & BE_3(b+x) & -CE_3(b+x) \\ -CE_3(b+x) & AE_3(b+x) & -BE_3(b+x) \\ BE_3(b+x) & CE_3(b+x) & AE_3(b+x) \end{bmatrix}$$

$$= \exp\left(\begin{bmatrix} 0 & b+x & 0 \\ 0 & 0 & -(b+x) \\ b+x & 0 & 0 \end{bmatrix}\right)$$

$$= \exp\left(\begin{bmatrix} 0 & b & 0 \\ 0 & 0 & -b \\ b & 0 & 0 \end{bmatrix}\right) \exp\left(\begin{bmatrix} 0 & x & 0 \\ 0 & 0 & -x \\ x & 0 & 0 \end{bmatrix}\right)$$

$$= \begin{bmatrix} AE_3(b) & BE_3(b) & -CE_3(b) \\ -CE_3(b) & AE_3(b) & -BE_3(b) \\ BE_3(b) & CE_3(b) & AE_3(b) \end{bmatrix} \begin{bmatrix} AE_3(x) & BE_3(x) & -CE_3(x) \\ -CE_3(x) & AE_3(x) & -BE_3(x) \\ BE_3(x) & CE_3(x) & AE_3(x) \end{bmatrix}$$

Thus, this matrix is a rotation matrix. By a similar procedure, the other matrix containing the E-type simple trigonometric functions is also a rotation matrix. Both procedures lead to the same trigonometric identities.

Trigonometric Identity 2, 3, & 4:

$$AE_3(b+x) = AE_3(b)AE_3(x) - BE_3(b)CE_3(x) - CE_3(b)BE_3(x)$$

$$BE_3(b+x) = AE_3(b)BE_3(x) + BE_3(b)AE_3(x) - CE_3(b)CE_3(x)$$

$$CE_3(b+x) = AE_3(b)CE_3(x) + BE_3(b)BE_3(x) + CE_3(b)AE_3(x)$$

The distance function is:

$$r = \sqrt[3]{a^3 - b^3 - c^3 - 3abc}$$

Note:

Since $e^a > 0$ for all values of a, then $r^3 > 0$ and so $x^3 - y^3 - z^3 - 3xyz > 0$.

The E-type simple trigonometric functions appear in the rotation matrix. Consideration of the series representations of the E-type simple trigonometric functions shows that they have a differentiation cycle (of length six).

We have a distance function, a set of two simple rotation matrices, and a set of three simple trigonometric functions. We have a natural space.

Again, this algebra is comprised of the set of matrices that equate to the polar form.

Summing the series of the simple trigonometric functions leads to:

$$AE_3(x) = \frac{1}{3}\left(e^{-x} + 2e^{\left(\frac{x}{2}\right)} \cos\left(\frac{\sqrt{3}}{2}x \right) \right)$$

$$BE_3(x) = \frac{1}{3}\left(-e^{-x} + e^{\left(\frac{x}{2}\right)}\left(\sqrt{3}\sin\left(\frac{\sqrt{3}}{2}x \right) + \cos\left(\frac{\sqrt{3}}{2}x \right) \right) \right)$$

$$CE_3(x) = \frac{1}{3}\left(e^{-x} + e^{\left(\frac{x}{2}\right)}\left(\sqrt{3}\sin\left(\frac{\sqrt{3}}{2}x \right) - \cos\left(\frac{\sqrt{3}}{2}x \right) \right) \right)$$

Section 7 : Chapter 3

The Polar Forms of the $C_3 L^1 E^1 H^1{}_{(j=-1,k=1)}$ and $C_3 L^1 E^1 H^1{}_{(j=-1,k=-1)}$ Algebras

We have: $C_3 L^1 E^1 H^1{}_{(j=-1,k=1)}$

$$\exp\left(\begin{bmatrix} a & b & c \\ -c & a & b \\ -b & -c & a \end{bmatrix}\right) =$$

$$\begin{bmatrix} h & 0 & 0 \\ 0 & h & 0 \\ 0 & 0 & h \end{bmatrix} \begin{bmatrix} AE_3(b) & BE_3(b) & CE_3(b) \\ -CE_3(b) & AE_3(b) & BE_3(b) \\ -BE_3(b) & -CE_3(b) & AE_3(b) \end{bmatrix} \begin{bmatrix} AH_3(c) & -CH_3(c) & BH_3(c) \\ -BH_3(c) & AH_3(c) & -CH_3(c) \\ CH_3(c) & -BH_3(c) & AH_3(c) \end{bmatrix}$$

and: $C_3 L^1 E^1 H^1{}_{(j=-1,k=-1)}$

$$\exp\left(\begin{bmatrix} a & b & c \\ -c & a & -b \\ -b & c & a \end{bmatrix}\right) =$$

$$\begin{bmatrix} h & 0 & 0 \\ 0 & h & 0 \\ 0 & 0 & h \end{bmatrix} \begin{bmatrix} AH_3(b) & BH_3(b) & -CH_3(b) \\ CH_3(b) & AH_3(b) & -BH_3(b) \\ -BH_3(b) & -CH_3(b) & AH_3(b) \end{bmatrix} \begin{bmatrix} AE_3(c) & CE_3(c) & BE_3(c) \\ -BE_3(c) & AE_3(c) & -CE_3(c) \\ -CE_3(c) & BE_3(c) & AE_3(c) \end{bmatrix}$$

The determinants of the rotation matrices being unity leads to the trigonometric identities previously found.

The distance functions are, respectively:

$$C_3 L^1 E^1 H^1{}_{(j=-1,k=1)} \quad : \quad r = \sqrt[3]{a^3 - b^3 + c^3 + 3abc}$$

$$C_3 L^1 E^1 H^1{}_{(j=-1,k=-1)} \quad : \quad r = \sqrt[3]{a^3 + b^3 - c^3 + 3abc}$$

Again, in each case, we have the requirements of a space, but these spaces are mixed spaces containing both H-type angles and E-Type angles.

Section 7 : Chapter 4

The 3-dimensional v-functions - an Introduction

The reader will doubtless have noticed that the 3-dimensional algebras have two simple rotation matrices in the polar form. What do we get if these two matrices are multiplied together? We have: $C_3 L^1 H^2_{(j=1,k=1)}$

$$\exp\left(\begin{bmatrix} a & b & c \\ c & a & b \\ b & c & a \end{bmatrix}\right) =$$

$$\begin{bmatrix} r & 0 & 0 \\ 0 & r & 0 \\ 0 & 0 & r \end{bmatrix} \begin{bmatrix} AH_3(b) & BH_3(b) & CH_3(b) \\ CH_3(b) & AH_3(b) & BH_3(b) \\ BH_3(b) & CH_3(b) & AH_3(b) \end{bmatrix} \begin{bmatrix} AH_3(c) & CH_3(c) & BH_3(c) \\ BH_3(c) & AH_3(c) & CH_3(c) \\ CH_3(c) & BH_3(c) & AH_3(c) \end{bmatrix}$$

We define:

$$v_{\left[C_3 L^1 H^2_{(j=1,k=1)}\right]} A(b,c) = AH_3(b)AH_3(c) + BH_3(b)BH_3(c) + CH_3(b)CH_3(c)$$

$$v_{\left[C_3 L^1 H^2_{(j=1,k=1)}\right]} B(b,c) = AH_3(c)BH_3(b) + AH_3(b)CH_3(c) + BH_3(c)CH_3(b)$$

$$v_{\left[C_3 L^1 H^2_{(j=1,k=1)}\right]} C(b,c) = AH_3(c)CH_3(b) + AH_3(b)BH_3(c) + BH_3(b)CH_3(c)$$

The subscripted bracket contains the name of the algebra, and A, B, C designate the different functions. The A, B, C correspond to the position in the cartesian algebraic matrix form. With this notation, the compound polar form is:

$$\exp\left(\begin{bmatrix} a & b & c \\ c & a & b \\ b & c & a \end{bmatrix}\right) =$$

$$\begin{bmatrix} r & 0 & 0 \\ 0 & r & 0 \\ 0 & 0 & r \end{bmatrix} \begin{bmatrix} v_{\left[C_3 L^1 H^2_{(j=1,k=1)}\right]} A(b,c) & v_{\left[C_3 L^1 H^2_{(j=1,k=1)}\right]} B(b,c) & v_{\left[C_3 L^1 H^2_{(j=1,k=1)}\right]} C(b,c) \\ v_{\left[C_3 L^1 H^2_{(j=1,k=1)}\right]} C(b,c) & v_{\left[C_3 L^1 H^2_{(j=1,k=1)}\right]} A(b,c) & v_{\left[C_3 L^1 H^2_{(j=1,k=1)}\right]} B(b,c) \\ v_{\left[C_3 L^1 H^2_{(j=1,k=1)}\right]} B(b,c) & v_{\left[C_3 L^1 H^2_{(j=1,k=1)}\right]} C(b,c) & v_{\left[C_3 L^1 H^2_{(j=1,k=1)}\right]} A(b,c) \end{bmatrix}$$

Clearly, the determinant of this matrix, being the product of two rotation matrices, is unity.

---

v-function Identity 1:

$$\left(v_{\left[C_3 L^1 H^2_{(j=1,k=1)}\right]} A(b,c)\right)^3 + \left(v_{\left[C_3 L^1 H^2_{(j=1,k=1)}\right]} B(b,c)\right)^3 + \left(v_{\left[C_3 L^1 H^2_{(j=1,k=1)}\right]} C(b,c)\right)^3$$

$$-3 v_{\left[C_3 L^1 H^2_{(j=1,k=1)}\right]} A(b,c) v_{\left[C_3 L^1 H^2_{(j=1,k=1)}\right]} B(b,c) v_{\left[C_3 L^1 H^2_{(j=1,k=1)}\right]} C(b,c) = 1$$

We have:

$$\begin{bmatrix} v_{\left[C_3 L^l H^2_{(j=1,k=1)}\right]}A(b+e,c+f) & v_{\left[C_3 L^l H^2_{(j=1,k=1)}\right]}B(b+e,c+f) & v_{\left[C_3 L^l H^2_{(j=1,k=1)}\right]}C(b+e,c+f) \\ v_{\left[C_3 L^l H^2_{(j=1,k=1)}\right]}C(b+e,c+f) & v_{\left[C_3 L^l H^2_{(j=1,k=1)}\right]}A(b+e,c+f) & v_{\left[C_3 L^l H^2_{(j=1,k=1)}\right]}B(b+e,c+f) \\ v_{\left[C_3 L^l H^2_{(j=1,k=1)}\right]}B(b+e,c+f) & v_{\left[C_3 L^l H^2_{(j=1,k=1)}\right]}C(b+e,c+f) & v_{\left[C_3 L^l H^2_{(j=1,k=1)}\right]}A(b+e,c+f) \end{bmatrix}$$

$$= \exp\left(\begin{bmatrix} 0 & b+e & c+f \\ c+f & 0 & b+e \\ b+e & c+f & 0 \end{bmatrix}\right)$$

$$= \exp\left(\begin{bmatrix} 0 & b & c \\ c & 0 & b \\ b & c & 0 \end{bmatrix}\right)\exp\left(\begin{bmatrix} 0 & e & f \\ f & 0 & e \\ e & f & 0 \end{bmatrix}\right)$$

$$= PROD\left(\begin{bmatrix} v_{\left[C_3 L^l H^2_{(j=1,k=1)}\right]}A(b,c) & v_{\left[C_3 L^l H^2_{(j=1,k=1)}\right]}B(b,c) & v_{\left[C_3 L^l H^2_{(j=1,k=1)}\right]}C(b,c) \\ v_{\left[C_3 L^l H^2_{(j=1,k=1)}\right]}C(b,c) & v_{\left[C_3 L^l H^2_{(j=1,k=1)}\right]}A(b,c) & v_{\left[C_3 L^l H^2_{(j=1,k=1)}\right]}B(b,c) \\ v_{\left[C_3 L^l H^2_{(j=1,k=1)}\right]}B(b,c) & v_{\left[C_3 L^l H^2_{(j=1,k=1)}\right]}C(b,c) & v_{\left[C_3 L^l H^2_{(j=1,k=1)}\right]}A(b,c) \end{bmatrix} \begin{bmatrix} v_{\left[C_3 L^l H^2_{(j=1,k=1)}\right]}A(e,f) & v_{\left[C_3 L^l H^2_{(j=1,k=1)}\right]}B(e,f) & v_{\left[C_3 L^l H^2_{(j=1,k=1)}\right]}C(e,f) \\ v_{\left[C_3 L^l H^2_{(j=1,k=1)}\right]}C(e,f) & v_{\left[C_3 L^l H^2_{(j=1,k=1)}\right]}A(e,f) & v_{\left[C_3 L^l H^2_{(j=1,k=1)}\right]}B(e,f) \\ v_{\left[C_3 L^l H^2_{(j=1,k=1)}\right]}B(e,f) & v_{\left[C_3 L^l H^2_{(j=1,k=1)}\right]}C(e,f) & v_{\left[C_3 L^l H^2_{(j=1,k=1)}\right]}A(e,f) \end{bmatrix}\right)$$

Leading to:

v-function Identity 2, 3, & 4:

$$v_{\left[C_3 L^l H^2_{(j=1,k=1)}\right]}A(b+e,c+f)=$$

$$v_{[\ ]}A(b,c)v_{[\ ]}A(e,f)+v_{[\ ]}B(b,c)v_{[\ ]}C(e,f)+v_{[\ ]}C(b,c)v_{[\ ]}B(e,f)$$

$$v_{\left[C_3 L^l H^2_{(j=1,k=1)}\right]}B(b+e,c+f)=$$

$$v_{[\ ]}B(b,c)v_{[\ ]}A(e,f)+v_{[\ ]}C(b,c)v_{[\ ]}C(e,f)+v_{[L]}A(b,c)v_{[\ ]}B(e,f)$$

$$v_{\left[C_3 L^l H^2_{(j=1,k=1)}\right]}C(b+e,c+f)=$$

$$v_{[\ ]}C(b,c)v_{[\ ]}A(e,f)+v_{[\ ]}A(b,c)v_{[\ ]}C(e,f)+v_{[\ ]}B(b,c)v_{[\ ]}B(e,f)$$

Thus the compound matrix has the properties necessary to be a rotation matrix.

Summing the series of the simple trigonometric functions leads to :

$$v_{\left[C_3 L^1 H^2_{(j=1,k=1)}\right]}A(b,c)=\frac{1}{3}\left(e^{(b+c)}+2e^{-\left(\frac{b+c}{2}\right)}\cos\left(\frac{\sqrt{3}}{2}(b-c)\right)\right)$$

$$v_{\left[C_3 L^1 H^2_{(j=1,k=1)}\right]}B(b,c)=\frac{1}{3}\left(e^{(b+c)}+e^{-\left(\frac{b+c}{2}\right)}\left(\sqrt{3}\sin\left(\frac{\sqrt{3}}{2}(b-c)\right)-\cos\left(\frac{\sqrt{3}}{2}(b-c)\right)\right)\right)$$

$$v_{\left[C_3 L^1 H^2_{(j=1,k=1)}\right]}C(b,c)=\frac{1}{3}\left(e^{(b+c)}-e^{-\left(\frac{b+c}{2}\right)}\left(\sqrt{3}\sin\left(\frac{\sqrt{3}}{2}(b-c)\right)+\cos\left(\frac{\sqrt{3}}{2}(b-c)\right)\right)\right)$$

The reader should compare these to the standard forms of the 3-dimensional simple trigonometric functions. These are not the simple trigonometric functions with a double argument. There is no misprinting of the signs of $\{(b+c),(b-c)\}$. In the case that one of the variables $\{b,c\}$ is zero, these functions will become the simple trigonometric functions of this algebra. We point out that:

$$v_{\left[C_3 L^1 H^2_{(j=1,k=1)}\right]}A(b,c)+v_{\left[C_3 L^1 H^2_{(j=1,k=1)}\right]}B(b,c)+v_{\left[LC_3 L^1 H^2_{(j=1,k=1)}\right]}C(b,c)=e^{(b+c)}$$

and so the v-functions are, at least in this sense, splittings of the exponential.

We have:

$$\begin{bmatrix} x & y & z \\ z & x & y \\ y & z & x \end{bmatrix}=\begin{bmatrix} r & 0 & 0 \\ 0 & r & 0 \\ 0 & 0 & r \end{bmatrix}\begin{bmatrix} v_{\left[C_3 L^1 H^2_{(j=1,k=1)}\right]}A(b,c) & v_{\left[C_3 L^1 H^2_{(j=1,k=1)}\right]}B(b,c) & v_{\left[C_3 L^1 H^2_{(j=1,k=1)}\right]}C(b,c) \\ v_{\left[C_3 L^1 H^2_{(j=1,k=1)}\right]}C(b,c) & v_{\left[C_3 L^1 H^2_{(j=1,k=1)}\right]}A(b,c) & v_{\left[C_3 L^1 H^2_{(j=1,k=1)}\right]}B(b,c) \\ v_{\left[C_3 L^1 H^2_{(j=1,k=1)}\right]}B(b,c) & v_{\left[C_3 L^1 H^2_{(j=1,k=1)}\right]}C(b,c) & v_{\left[C_3 L^1 H^2_{(j=1,k=1)}\right]}A(b,c) \end{bmatrix}$$

Normalising this brings us on to the unit sphere:

$$\frac{1}{\sqrt[3]{x^3+y^3+z^3-3xyz}}\begin{bmatrix} x & y & z \\ z & x & y \\ y & z & x \end{bmatrix}=$$

$$\begin{bmatrix} v_{\left[C_3 L^1 H^2_{(j=1,k=1)}\right]}A(b,c) & v_{\left[C_3 L^1 H^2_{(j=1,k=1)}\right]}B(b,c) & v_{\left[C_3 L^1 H^2_{(j=1,k=1)}\right]}C(b,c) \\ v_{\left[C_3 L^1 H^2_{(j=1,k=1)}\right]}C(b,c) & v_{\left[C_3 L^1 H^2_{(j=1,k=1)}\right]}A(b,c) & v_{\left[C_3 L^1 H^2_{(j=1,k=1)}\right]}B(b,c) \\ v_{\left[C_3 L^1 H^2_{(j=1,k=1)}\right]}B(b,c) & v_{\left[C_3 L^1 H^2_{(j=1,k=1)}\right]}C(b,c) & v_{\left[C_3 L^1 H^2_{(j=1,k=1)}\right]}A(b,c) \end{bmatrix}$$

Leading to:

$$\frac{x}{\sqrt[3]{x^3 + y^3 + z^3 - 3xyz}} = v_{[\,]}A$$

$$\frac{y}{\sqrt[3]{x^3 + y^3 + z^3 - 3xyz}} = v_{[\,]}B$$

$$\frac{z}{\sqrt[3]{x^3 + y^3 + z^3 - 3xyz}} = v_{[\,]}C$$

And so, these are functions are projection functions from the unit sphere on to the axes.

This algebraic field is the set of matrices of the form:

$$\begin{bmatrix} x & y & z \\ z & x & y \\ y & z & x \end{bmatrix} : \begin{cases} x = e^a \dfrac{1}{3}\left(e^{(b+c)} + 2e^{-\left(\frac{b+c}{2}\right)}\cos\left(\dfrac{\sqrt{3}}{2}(b-c)\right)\right) \\[2mm] y = e^a \dfrac{1}{3}\left(e^{-(b+c)} + e^{-\left(\frac{b+c}{2}\right)}\left(\sqrt{3}\sin\left(\dfrac{\sqrt{3}}{2}(b-c)\right) - \cos\left(\dfrac{\sqrt{3}}{2}(b-c)\right)\right)\right) \\[2mm] z = e^a \dfrac{1}{3}\left(e^{(b+c)} - e^{-\left(\frac{b+c}{2}\right)}\left(\sqrt{3}\sin\left(\dfrac{\sqrt{3}}{2}(b-c)\right) + \cos\left(\dfrac{\sqrt{3}}{2}(b-c)\right)\right)\right) \end{cases}$$

for unrestricted (real) values of $\{a,b,c\}$.

By similar means, we can define v-functions for the other algebras.

The $C_3 L^1 E^2{}_{(j=1,k=-1)}$ algebra:

$$v_{\left[C_3 L^1 E^2{}_{(j=1,k=-1)}\right]}A(b,c) = \frac{1}{3}\left(e^{-(b+c)} + 2e^{\left(\frac{b+c}{2}\right)}\cos\left(\dfrac{\sqrt{3}}{2}(b-c)\right)\right)$$

$$v_{\left[C_3 L^1 E^2{}_{(j=1,k=-1)}\right]}B(b,c) = \frac{1}{3}\left(-e^{-(b+c)} + e^{\left(\frac{b+c}{2}\right)}\left(\sqrt{3}\sin\left(\dfrac{\sqrt{3}}{2}(b-c)\right) + \cos\left(\dfrac{\sqrt{3}}{2}(b-c)\right)\right)\right)$$

$$v_{\left[C_3 L^1 E^2{}_{(j=1,k=-1)}\right]}C(b,c) = \frac{1}{3}\left(-e^{-(b+c)} - e^{\left(\frac{b+c}{2}\right)}\left(\sqrt{3}\sin\left(\dfrac{\sqrt{3}}{2}(b-c)\right) - \cos\left(\dfrac{\sqrt{3}}{2}(b-c)\right)\right)\right)$$

The $C_3 L^1 E^1 H^1_{(j=-1,k=1)}$ algebra:

$$V_{\left[C_3 L^1 E^1 H^1_{(j=-1,k=1)}\right]} A(b,c) = \frac{1}{3}\left(e^{-(b-c)} + 2e^{\left(\frac{b-c}{2}\right)}\cos\left(\frac{\sqrt{3}}{2}(b+c)\right)\right)$$

$$V_{\left[C_3 L^1 E^1 H^1_{(j=-1,k=1)}\right]} B(b,c) = \frac{1}{3}\left(-e^{-(b-c)} + e^{\left(\frac{b-c}{2}\right)}\left(\sqrt{3}\sin\left(\frac{\sqrt{3}}{2}(b+c)\right) + \cos\left(\frac{\sqrt{3}}{2}(b+c)\right)\right)\right)$$

$$V_{\left[C_3 L^1 E^1 H^1_{(j=-1,k=1)}\right]} C(b,c) = \frac{1}{3}\left(e^{-(b-c)} + e^{\left(\frac{b-c}{2}\right)}\left(\sqrt{3}\sin\left(\frac{\sqrt{3}}{2}(b+c)\right) - \cos\left(\frac{\sqrt{3}}{2}(b+c)\right)\right)\right)$$

The $C_3 L^1 E^1 H^1_{(j=-1,k=-1)}$ algebra:

$$V_{\left[C_3 L^1 E^1 H^1_{(j=-1,k=-1)}\right]} A(b,c) = \frac{1}{3}\left(e^{(b-c)} + 2e^{-\left(\frac{b-c}{2}\right)}\cos\left(\frac{\sqrt{3}}{2}(b+c)\right)\right)$$

$$V_{\left[C_3 L^1 E^1 H^1_{(j=-1,k=-1)}\right]} B(b,c) = \frac{1}{3}\left(e^{(b-c)} + e^{-\left(\frac{b-c}{2}\right)}\left(\sqrt{3}\sin\left(\frac{\sqrt{3}}{2}(b+c)\right) - \cos\left(\frac{\sqrt{3}}{2}(b+c)\right)\right)\right)$$

$$V_{\left[C_3 L^1 E^1 H^1_{(j=-1,k=-1)}\right]} C(b,c) = \frac{1}{3}\left(-e^{(b-c)} + e^{-\left(\frac{b-c}{2}\right)}\left(\sqrt{3}\sin\left(\frac{\sqrt{3}}{2}(b+c)\right) + \cos\left(\frac{\sqrt{3}}{2}(b+c)\right)\right)\right)$$

Section 7 : Chapter 5

The General Polar Form

The 3-dimensional general polar form is:

$$\begin{bmatrix} e^a & 0 & 0 \\ 0 & e^a & 0 \\ 0 & 0 & e^a \end{bmatrix}$$

$$\begin{bmatrix} AH_3\left((jk)^{\frac{1}{3}}b\right) & \dfrac{1}{(jk)^{\frac{1}{3}}}BH_3\left((jk)^{\frac{1}{3}}b\right) & \dfrac{k}{(jk)^{\frac{2}{3}}}CH_3\left((jk)^{\frac{1}{3}}b\right) \\[2em] j\left(\dfrac{k}{(jk)^{\frac{2}{3}}}CH_3\left((jk)^{\frac{1}{3}}b\right)\right) & AH_3\left((jk)^{\frac{1}{3}}b\right) & k\left(\dfrac{1}{(jk)^{\frac{1}{3}}}BH_3\left((jk)^{\frac{1}{3}}b\right)\right) \\[2em] j\left(\dfrac{1}{(jk)^{\frac{1}{3}}}BH_3\left((jk)^{\frac{1}{3}}b\right)\right) & \dfrac{j}{k}\left(\dfrac{k}{(jk)^{\frac{2}{3}}}CH_3\left((jk)^{\frac{1}{3}}b\right)\right) & AH_3\left((jk)^{\frac{1}{3}}b\right) \end{bmatrix}$$

$$\begin{bmatrix} AH_3\left(\left(\dfrac{j^2}{k}\right)^{\frac{1}{3}}c\right) & \dfrac{j}{k}\left(\dfrac{k}{j^2}\right)^{\frac{2}{3}}CH_3\left(\left(\dfrac{j^2}{k}\right)^{\frac{1}{3}}c\right) & \left(\dfrac{k}{j^2}\right)^{\frac{1}{3}}BH_3\left(\left(\dfrac{j^2}{k}\right)^{\frac{1}{3}}c\right) \\[2em] j\left(\left(\dfrac{k}{j^2}\right)^{\frac{1}{3}}BH_3\left(\left(\dfrac{j^2}{k}\right)^{\frac{1}{3}}c\right)\right) & AH_3\left(\left(\dfrac{j^2}{k}\right)^{\frac{1}{3}}c\right) & k\dfrac{j}{k}\left(\dfrac{k}{j^2}\right)^{\frac{2}{3}}CH_3\left(\left(\dfrac{j^2}{k}\right)^{\frac{1}{3}}c\right) \\[2em] j\dfrac{j}{k}\left(\dfrac{k}{j^2}\right)^{\frac{2}{3}}CH_3\left(\left(\dfrac{j^2}{k}\right)^{\frac{1}{3}}c\right) & \dfrac{j}{k}\left(\dfrac{k}{j^2}\right)^{\frac{1}{3}}BH_3\left(\left(\dfrac{j^2}{k}\right)^{\frac{1}{3}}c\right) & AH_3\left(\left(\dfrac{j^2}{k}\right)^{\frac{1}{3}}c\right) \end{bmatrix}$$

The determinants of the rotation matrices, in both cases, are:

$$AH_3^{\,3}(\ )+BH_3^{\,3}(\ )+CH_3^{\,3}(\ )-3AH_3(\ )BH_3(\ )CH_3(\ )=1$$

The distance function is:

$$r = \sqrt[3]{a^3 + jkb^3 + \frac{j^2}{k}c^3 - 3jabc}$$

When $\{j=1,k=-1\}$, this general form becomes the $C_3L^1E^2_{(j=1,k=-1)}$ algebra.

We equate the b rotation matrices of these two cases.

$$
\begin{bmatrix}
AH_3\left((-1)^{\frac{1}{3}}b\right) & \dfrac{1}{(-1)^{\frac{1}{3}}}BH_3\left((-1)^{\frac{1}{3}}b\right) & \dfrac{-1}{(-1)^{\frac{2}{3}}}CH_3\left((-1)^{\frac{1}{3}}b\right) \\[3mm]
\left(\dfrac{-1}{(-1)^{\frac{2}{3}}}CH_3\left((-1)^{\frac{1}{3}}b\right)\right) & AH_3\left((-1)^{\frac{1}{3}}b\right) & -\left(\dfrac{1}{(-1)^{\frac{1}{3}}}BH_3\left((-1)^{\frac{1}{3}}b\right)\right) \\[3mm]
\left(\dfrac{1}{(-1)^{\frac{1}{3}}}BH_3\left((-1)^{\frac{1}{3}}b\right)\right) & -\left(\dfrac{-1}{(-1)^{\frac{2}{3}}}CH_3\left((-1)^{\frac{1}{3}}b\right)\right) & AH_3\left((-1)^{\frac{1}{3}}b\right)
\end{bmatrix}
$$

$$
=\begin{bmatrix}
AE_3(b) & BE_3(b) & -CE_3(b) \\
-CE_3(b) & AE_3(b) & -BE_3(b) \\
BE_3(b) & CE_3(b) & AE_3(b)
\end{bmatrix}
$$

Leading to:

$$
AH_3\left((-1)^{\frac{1}{3}}b\right)=AE_3(b)
$$

$$
BH_3\left((-1)^{\frac{1}{3}}b\right)=(-1)^{\frac{1}{3}}BE_3(b)
$$

$$
CH_3\left((-1)^{\frac{1}{3}}b\right)=(1)^{\frac{1}{3}}CE_3(b)
$$

which is analogous to the 2-dimensional case relating the euclidean trigonometric functions to the hyperbolic trigonometric functions. The reader should remember that, within these algebras, there is no $\hat{i}=\sqrt[2]{-1}$ and that $\left\{(-1)^{\frac{1}{3}},(1)^{\frac{1}{3}}\right\}$ are not euclidean complex numbers. The same results can be derived from any of the other algebras previously discussed.

Section 7 : Chapter 6

Differentiation and the Cauchy-Riemann Equations in 3-dimensions

Consider the 3 by 3 matrix function:

$$C_3 L^1 H^2_{(j=1,k=1)} : f\left(\begin{bmatrix} a & b & c \\ c & a & b \\ b & c & a \end{bmatrix}\right) \rightarrow \begin{bmatrix} u(a,b,c) & v(a,b,c) & w(a,b,c) \\ w(a,b,c) & u(a,b,c) & v(a,b,c) \\ v(a,b,c) & w(a,b,c) & u(a,b,c) \end{bmatrix}$$

We have:

$$\frac{\partial\left(\begin{bmatrix} u(a,b,c) & v(a,b,c) & w(a,b,c) \\ w(a,b,c) & u(a,b,c) & v(a,b,c) \\ v(a,b,c) & w(a,b,c) & u(a,b,c) \end{bmatrix}\right)}{\partial\begin{bmatrix} a & 0 & 0 \\ 0 & a & 0 \\ 0 & 0 & a \end{bmatrix}} = \begin{bmatrix} \frac{\partial u}{\partial a} & \frac{\partial v}{\partial a} & \frac{\partial w}{\partial a} \\ \frac{\partial w}{\partial a} & \frac{\partial u}{\partial a} & \frac{\partial v}{\partial a} \\ \frac{\partial v}{\partial a} & \frac{\partial w}{\partial a} & \frac{\partial u}{\partial a} \end{bmatrix}$$

$$\frac{\partial\left(\begin{bmatrix} u(a,b,c) & v(a,b,c) & w(a,b,c) \\ w(a,b,c) & u(a,b,c) & v(a,b,c) \\ v(a,b,c) & w(a,b,c) & u(a,b,c) \end{bmatrix}\right)}{\partial\begin{bmatrix} 0 & b & 0 \\ 0 & 0 & b \\ b & 0 & 0 \end{bmatrix}} = \begin{bmatrix} \frac{\partial v}{\partial b} & \frac{\partial w}{\partial b} & \frac{\partial u}{\partial b} \\ \frac{\partial u}{\partial b} & \frac{\partial v}{\partial b} & \frac{\partial w}{\partial b} \\ \frac{\partial w}{\partial b} & \frac{\partial u}{\partial b} & \frac{\partial v}{\partial b} \end{bmatrix}$$

$$\frac{\partial\left(\begin{bmatrix} u(a,b,c) & v(a,b,c) & w(a,b,c) \\ w(a,b,c) & u(a,b,c) & v(a,b,c) \\ v(a,b,c) & w(a,b,c) & u(a,b,c) \end{bmatrix}\right)}{\partial\begin{bmatrix} 0 & 0 & c \\ c & 0 & 0 \\ 0 & c & 0 \end{bmatrix}} = \begin{bmatrix} \frac{\partial w}{\partial c} & \frac{\partial u}{\partial c} & \frac{\partial v}{\partial c} \\ \frac{\partial v}{\partial c} & \frac{\partial w}{\partial c} & \frac{\partial u}{\partial c} \\ \frac{\partial u}{\partial c} & \frac{\partial v}{\partial c} & \frac{\partial w}{\partial c} \end{bmatrix}$$

Leading to the Cauchy Riemann equations for the $C_3 L^1 H^2_{(j=1,k=1)}$ algebra:

$$\frac{\partial u}{\partial a} = \frac{\partial v}{\partial b} = \frac{\partial w}{\partial c}$$

$$\frac{\partial v}{\partial a} = \frac{\partial w}{\partial b} = \frac{\partial u}{\partial c}$$

$$\frac{\partial w}{\partial a} = \frac{\partial u}{\partial b} = \frac{\partial v}{\partial c}$$

Section 7 : Chapter 7

Non-Matrix Notation of the Algebras

Although we are using matrix notation to do our algebras, it is sometimes useful and time saving to revert to non-matrix notation. We find the non-matrix notation by inspection of the matrix notation. We demonstrate with the euclidean complex numbers and a 3-dimensional algebra.

Addition: In all the natural algebras, addition is component-wise.

Euclidean Complex Numbers:
This is a 2-dimensional algebra. In non-matrix notation, we can denote one axis by the absence of any symbol but we need a symbol to denote the other axis. We choose a symbol with a hat on, $\hat{i}$.

Multiplication: We obtain the multiplication operation by inspection of the top row of the product matrix.

$$\begin{bmatrix} a & b \\ -b & a \end{bmatrix}\begin{bmatrix} c & d \\ -d & c \end{bmatrix} = \begin{bmatrix} ac-bd & ad+bc \\ & \end{bmatrix}$$

$$\left(a+\hat{i}b\right)\left(c+\hat{i}d\right) = \left(ac-bd\right)+\hat{i}\left(ad+bc\right)$$

Conjugation: The conjugate of an algebraic matrix form is the adjoint matrix of that algebraic matrix form. The conjugate is obtained from the top row of the adjoint matrix.

$$adj\left(\begin{bmatrix} a & b \\ -b & a \end{bmatrix}\right) = \begin{bmatrix} a & -b \\ b & a \end{bmatrix}$$

$$\overline{\left(a+\hat{i}b\right)} = a-\hat{i}b$$

Since the adjoint matrix is the inverse matrix multiplied by the determinant, the product of the matrix and its conjugate is the determinant. We use this to discover the nature of the $\hat{i}$. To do this, we multiply the number and its conjugate together as if the variables were all real numbers; we do not use the multiplication operation taken from the matrix.

$$\left(a+\hat{i}b\right)\left(a-\hat{i}b\right) = a^2 - a\hat{i}b + \hat{i}ba - \hat{i}^2 b^2 = a^2 - \hat{i}^2 b^2$$

Equating this to the determinant of the algebraic matrix form, $a^2 + b^2$, leads to

$$\hat{i}^2 = -1.^{61}$$

---

[61] We could achieve the same by comparing the two products.

The 3-dimensional Case:

These are 3-dimensional algebras. One of the dimensions is the real numbers; to denote the other two dimensions, we need two symbols with hats on, $\hat{p}, \hat{q}$. We use the same two symbols in each of the four flat algebras. Since these symbols mean different things in each of the different flat algebras, this is a trap for the unwary. However, we feel it is preferable to the alternative option of confusing the wary with a plethora of symbols.

Addition is component-wise.

$$\left(a + \hat{p}b + \hat{q}c\right) + \left(d + \hat{p}e + \hat{q}f\right) = (a+d) + \hat{p}(b+e) + \hat{q}(c+f)$$

Multiplication:

$$\begin{bmatrix} a & b & c \\ jc & a & kb \\ jb & \left(\dfrac{j}{k}\right)c & a \end{bmatrix} \begin{bmatrix} d & e & f \\ jf & d & ke \\ je & \left(\dfrac{j}{k}\right)f & d \end{bmatrix} = \begin{bmatrix} ad + jbf + jce & ae+bd+\left(\dfrac{j}{k}\right)cf & af+kbe+cd \\ \sim & \sim & \sim \\ \sim & \sim & \sim \end{bmatrix}$$

$$\left(a + \hat{p}b + \hat{q}c\right)\left(d + \hat{p}e + \hat{q}f\right) = (ad + jbf + jce) + \hat{p}\left(ae+bd+\left(\dfrac{j}{k}\right)cf\right) + \hat{q}(af+kbe+cd)$$

Conjugation:

$$adj\left(\begin{bmatrix} a & b & c \\ jc & a & kb \\ jb & \left(\dfrac{j}{k}\right)c & a \end{bmatrix}\right) = \begin{bmatrix} a^2 - jbc & \left(\dfrac{j}{k}\right)c^2 - ab & kb^2 - ac \\ \sim & \sim & \sim \\ \sim & \sim & \sim \end{bmatrix}$$

$$\overline{\left(a + \hat{p}b + \hat{q}c\right)} = \left(a^2 - jbc\right) + \hat{p}\left(\left(\dfrac{j}{k}\right)c^2 - ab\right) + \hat{q}(kb^2 - ac)$$

The product of the 3-dimensional complex number and its conjugate is the determinant. We multiply the two together as if the variables were real numbers.

$$\left(a + \hat{p}b + \hat{q}c\right)\left(\left(a^2 - jbc\right) + \hat{p}\left(\left(\dfrac{j}{k}\right)c^2 - ab\right) + \hat{q}(kb^2 - ac)\right)$$

$$= \left(a^3 - jabc\right) + \hat{p}\left(\left(\dfrac{j}{k}\right)ac^2 - a^2b\right) + \hat{q}(kab^2 - a^2c)$$

$$+ \hat{p}\left(a^2b - jb^2c\right) + \hat{p}^2\left(\left(\dfrac{j}{k}\right)bc^2 - ab^2\right) + \hat{p}\hat{q}(kb^3 - abc)$$

$$+ \hat{q}\left(a^2c - jbc^2\right) + \hat{q}\hat{p}\left(\left(\dfrac{j}{k}\right)c^3 - abc\right) + \hat{q}^2(kb^2c - ac^2)$$

Leading to:

$$a^3 + \hat{p}\hat{q}kb^3 + \hat{q}\hat{p}\left(\frac{j}{k}\right)c^3 + abc\left(-j - \hat{p}\hat{q} - \hat{q}\hat{p}\right)$$

$$+\hat{p}\left(\left(\frac{j}{k}\right)ac^2 - jb^2c\right) + \hat{q}\left(kab^2 - jbc^2\right)$$

$$+\hat{p}^2\left(\left(\frac{j}{k}\right)bc^2 - ab^2\right) + \hat{q}^2\left(kb^2c - ac^2\right)$$

There are four different flat algebras depending upon the values of $\{j,k\}$, but all four flat algebras have determinants with the four terms $\{a^3, b^3, c^3, 3abc\}$; only the signs vary. We take them case-by-case and compare the above product to the determinant. We get:

Case: $j = k = +1$:

$$a^3 + \hat{p}\hat{q}b^3 + \hat{q}\hat{p}c^3 + abc\left(-1 - \hat{p}\hat{q} - \hat{q}\hat{p}\right)$$

$$+\hat{p}\left(ac^2 - b^2c\right) + \hat{q}\left(ab^2 - bc^2\right)$$

$$+\hat{p}^2\left(bc^2 - ab^2\right) + \hat{q}^2\left(b^2c - ac^2\right)$$

$\det\left(C_3 L^1 H^2_{(j=1,k=1)}\right) = a^3 + b^3 + c^3 - 3abc$ which compels us to conclude that

$$\left\{\left\{\hat{q}\hat{p} = \hat{p}\hat{q} = +1, \hat{p}^2 = \hat{q}, \hat{q}^2 = \hat{p}\right\} \Rightarrow \left\{\hat{p}^3 = +1, \hat{q}^3 = +1\right\}\right\}$$

The other flat cases lead to:

$$C_3 L^1 E^2_{(j=1,k=-1)} : \left\{\hat{p}\hat{q} = \hat{q}\hat{p} = +1, \hat{q}^2 = -\hat{p}, \hat{p}^2 = -\hat{q}\right\}\left\{\hat{p}^3 = -1, \hat{q}^3 = -1\right\}$$

$$C_3 L^1 E^1 H^1_{(j=-1,k=1)} : \left\{\hat{p}\hat{q} = \hat{q}\hat{p} = -1, \hat{q}^2 = -\hat{p}, \hat{p}^2 = +\hat{q}\right\}\left\{\hat{p}^3 = -1, \hat{q}^3 = +1\right\}$$

$$C_3 L^1 E^1 H^1_{(j=-1,k=-1)} : \left\{\hat{p}\hat{q} = \hat{q}\hat{p} = -1, \hat{q}^2 = +\hat{p}, \hat{p}^2 = -\hat{q}\right\}\left\{\hat{p}^3 = +1, \hat{q}^3 = -1\right\}$$

Aside:
We see that the 3-dimensional algebras are associated with the cube roots of ± 1. It is the case in general that the n-dimensional algebras are associated with the nth roots of ± 1, as will become clear later.

In the case $\left\{\hat{p}^2 = \hat{q}, \hat{p}^3 = +1\right\}$, we have:

$$e^{\hat{p}\theta} = 1 + \hat{p}\theta + \frac{\hat{p}^2\theta^2}{2!} + \frac{\hat{p}^3\theta^3}{3!} + \frac{\hat{p}^4\theta^4}{4!} + \frac{\hat{p}^5\theta^5}{5!} + \dots$$

$$= 1 + \hat{p}\theta + \frac{\hat{q}\theta^2}{2!} + \frac{\theta^3}{3!} + \frac{\hat{p}\theta^4}{4!} + \frac{\hat{q}\theta^5}{5!} + \dots$$

$$= \left(1 + \frac{\theta^3}{3!} + \dots\right) + \hat{p}\left(\theta + \frac{\theta^4}{4!} + \dots\right) + \hat{q}\left(\frac{\theta^2}{2!} + \frac{\theta^5}{5!} + \dots\right)$$

$$= AH_3(\theta) + \hat{p}BH_3(\theta) + \hat{q}CH_3(\theta)$$

Similarly:

$$e^{\hat{q}\theta} = AH_3(\theta) + \hat{q}BH_3(\theta) + \hat{p}CH_3(\theta)$$

This is analogous to the 2-dimensional case[62]:

$$e^{\hat{i}\theta} = \cos\theta + \hat{i}\sin\theta$$

Clearly, there are similar identities for the other 3-dimensional flat algebras. Obscurely, there are similar identities for all the higher dimensional flat algebras.

It follows that:

$$e^{\hat{p}\frac{\theta}{n}} = AH_3\left(\frac{\theta}{n}\right) + \hat{p}BH_3\left(\frac{\theta}{n}\right) + \hat{q}CH_3\left(\frac{\theta}{n}\right)$$

and:

$$e^{\hat{q}\frac{\theta}{n}} = AH_3\left(\frac{\theta}{n}\right) + \hat{q}BH_3\left(\frac{\theta}{n}\right) + \hat{p}CH_3\left(\frac{\theta}{n}\right)$$

And:

$$e^{\left(b\hat{p}+c\hat{q}\right)} = e^{b\hat{p}}e^{c\hat{q}}$$

$$= \left(AH_3(b) + \hat{p}BH_3(b) + \hat{q}CH_3(b)\right)\left(AH_3(c) + \hat{p}CH_3(c) + \hat{q}BH_3(c)\right)$$

$$= AH_3(b)AH_3(c) + \hat{p}AH_3(b)CH_3(c) + \hat{q}AH_3(b)BH_3(c)$$

$$+ \hat{p}BH_3(b)AH_3(c) + \hat{q}BH_3(b)CH_3(c) + BH_3(b)BH_3(c)$$

$$+ \hat{q}CH_3(b)AH_3(c) + CH_3(b)CH_3(c) + \hat{p}CH_3(b)BH_3(c)$$

$$= v_{\left[C_3L^1H^2_{(j=1,k=1)}\right]}A(b,c) + \hat{p}.v_{\left[C_3L^1H^2_{(j=1,k=1)}\right]}B(b,c) + \hat{q}.v_{\left[C_3L^1H^2_{(j=1,k=1)}\right]}C(b,c)$$

It follows that:

$$e^{\left(\frac{a+b\hat{p}+c\hat{q}}{n}\right)} = e^{\frac{a}{n}}\left(v_{\left[C_3L^1H^2_{(j=1,k=1)}\right]}A\left(\frac{b}{n},\frac{c}{n}\right) + \hat{p}.v_{\left[C_3L^1H^2_{(j=1,k=1)}\right]}B\left(\frac{b}{n},\frac{c}{n}\right) + \hat{q}.v_{\left[C_3L^1H^2_{(j=1,k=1)}\right]}C\left(\frac{b}{n},\frac{c}{n}\right)\right)$$

Hm! perhaps the simple trigonometric functions and the v-functions really are trigonometric functions after all!

---

[62] In 2005, physicists voted this formula the 4th most important formula in the known universe.

Section 7 : Chapter 8

The 2-dimensional Shadows of the 3-dimensional Algebras

In non-matrix notation, the $C_3 L^1 H^2{}_{(j=1,k=1)}$ algebra is the algebra of $a + b\hat{p} + c\hat{q}$ with the multiplicative relations:

$$\left\{ \hat{q}\hat{p} = \hat{p}\hat{q} = +1, \hat{p}^2 = \hat{q}, \hat{q}^2 = \hat{p}, \hat{p}^3 = +1, \hat{q}^3 = +1 \right\}$$

In the euclidean complex numbers, the cube roots of unity have the same multiplicative relations with:

$$\left\{ \hat{p} \equiv -\frac{1}{2} + \frac{\sqrt{3}}{2}\hat{i}, \hat{q} \equiv -\frac{1}{2} - \frac{\sqrt{3}}{2}\hat{i} \right\}$$

If we substitute these complex roots into the non-matrix representation of the $C_3 L^1 H^2{}_{(j=1,k=1)}$ algebra, we get:

$$a + b\hat{p} + c\hat{q} = a + b\left(-\frac{1}{2} + \frac{\sqrt{3}}{2}\hat{i} \right) + c\left(-\frac{1}{2} - \frac{\sqrt{3}}{2}\hat{i} \right)$$

$$= \left(a - \frac{b+c}{2} \right) + \hat{i}\left(\frac{\sqrt{3}}{2}(b-c) \right)$$

When we reduce the $C_3 L^1 H^2{}_{(j=1,k=1)}$ algebra to the euclidean complex numbers, we are squeezing a 3-dimensional algebra into a 2-dimensional representation of that algebra. We say that the 2-dimensional representation is a 2-dimensional shadow of the 3-dimensional algebra.

Nomenclature:

Shadow Algebra

A shadow algebra is a representation of an algebra in the euclidean complex numbers.

We still have three independent variables $\{a, b, c\}$ in the shadow of the $C_3 L^1 H^2{}_{(j=1,k=1)}$ algebra. The euclidean shadow is an algebraic field for all values of the three variables – it is the euclidean complex numbers. All natural algebraic fields are associated with a natural space. Thus, we are shadowing the 3-dimensional space of the $C_3 L^1 H^2{}_{(j=1,k=1)}$ algebra in the 2-dimensional space of the euclidean complex numbers.

Physical 2-dimensional shadows of 3-dimensional objects retain some of the detail of the 3-dimensional object but also lose some detail. So it is with shadow algebras and shadow spaces. The distance function of the $C_3 L^1 H^2{}_{(j=1,k=1)}$ algebra is based on the determinant of the matrix form of that algebra, which is:

$$a^3 + b^3 + c^3 - 3abc = (a + b + c)(a^2 + b^2 + c^2 - ab - ac - bc)$$

The shadow algebra, being a natural algebra, has, within it, a distance function based on the determinant, which is:

$$\left(a-\frac{(b+c)}{2}\right)^2 + \left(\frac{\sqrt{3}}{2}(b-c)\right)^2 = \left(a^2+b^2+c^2-ab-ac-bc\right)$$

We have lost the factor $(a+b+c)$. This factor is of the form of the 1-dimensional metric; we seem to have squashed out one dimension.

The reader should realise that we are not cutting a 2-dimensional sub-space out of the 3-dimensional space[63]; we are representing the whole of the 3-dimensional space as a 2-dimensional shadow of itself.

The polar form of the shadow representation is:

$$\exp\left(\begin{bmatrix} a-\dfrac{(b+c)}{2} & \dfrac{\sqrt{3}}{2}(b-c) \\ -\dfrac{\sqrt{3}}{2}(b-c) & a-\dfrac{(b+c)}{2} \end{bmatrix}\right)$$

$$= \exp\left(\begin{bmatrix} a & 0 \\ 0 & a \end{bmatrix} + \begin{bmatrix} -\dfrac{(b+c)}{2} & 0 \\ 0 & -\dfrac{(b+c)}{2} \end{bmatrix} + \begin{bmatrix} 0 & \dfrac{\sqrt{3}}{2}(b-c) \\ -\dfrac{\sqrt{3}}{2}(b-c) & 0 \end{bmatrix}\right)$$

$$= \begin{bmatrix} e^a & 0 \\ 0 & e^a \end{bmatrix} \begin{bmatrix} e^{-\frac{(b+c)}{2}} & 0 \\ 0 & e^{-\frac{(b+c)}{2}} \end{bmatrix} \begin{bmatrix} \cos\left(\dfrac{\sqrt{3}}{2}(b-c)\right) & \sin\left(\dfrac{\sqrt{3}}{2}(b-c)\right) \\ -\sin\left(\dfrac{\sqrt{3}}{2}(b-c)\right) & \cos\left(\dfrac{\sqrt{3}}{2}(b-c)\right) \end{bmatrix}$$

We see that the length within the 3-dimensional algebra, e^a, has now become $e^{a-\frac{b+c}{2}}$. The 3-dimensional algebras have two rotation matrices in the simple polar form. When we combine these together to form the compound polar form, we get a single rotation matrix containing the v-functions. The reader is urged to compare the above shadow with the v-functions for the $C_3L^lH^2_{(j=1,k=1)}$ algebra.

Analogous results occur for the three other 3-dimensional algebras.

---

[63] 3-dimensional natural spaces do not have 2-dimensional sub-spaces.

Section 7 : Chapter 9

Logarithms - $C_3 L^1 H^2_{(j=1,k=1)}$

The simple polar form of the $C_3 L^1 H^2_{(j=1,k=1)}$ algebra is:

$$\begin{bmatrix} a & b & c \\ c & a & b \\ b & c & a \end{bmatrix} = PROD \left(\begin{bmatrix} h & 0 & 0 \\ 0 & h & 0 \\ 0 & 0 & h \end{bmatrix} \begin{bmatrix} AH_3(\beta) & BH_3(\beta) & CH_3(\beta) \\ CH_3(\beta) & AH_3(\beta) & BH_3(\beta) \\ BH_3(\beta) & CH_3(\beta) & AH_3(\beta) \end{bmatrix} \begin{bmatrix} AH_3(\chi) & CH_3(\chi) & BH_3(\chi) \\ BH_3(\chi) & AH_3(\chi) & CH_3(\chi) \\ CH_3(\chi) & BH_3(\chi) & AH_3(\chi) \end{bmatrix} \right)$$

Thus:

$$\log m_{[e]} \left(\begin{bmatrix} a & b & c \\ c & a & b \\ b & c & a \end{bmatrix} \right)$$

$$= \log m_{[e]} \left(\begin{bmatrix} h & 0 & 0 \\ 0 & h & 0 \\ 0 & 0 & h \end{bmatrix} \begin{bmatrix} AH_3(\beta) & BH_3(\beta) & CH_3(\beta) \\ CH_3(\beta) & AH_3(\beta) & BH_3(\beta) \\ BH_3(\beta) & CH_3(\beta) & AH_3(\beta) \end{bmatrix} \begin{bmatrix} AH_3(\chi) & CH_3(\chi) & BH_3(\chi) \\ BH_3(\chi) & AH_3(\chi) & CH_3(\chi) \\ CH_3(\chi) & BH_3(\chi) & AH_3(\chi) \end{bmatrix} \right)$$

$$= \log m_{[e]} \left(\begin{bmatrix} h & 0 & 0 \\ 0 & h & 0 \\ 0 & 0 & h \end{bmatrix} \right) + \log m_{[e]} \left(\begin{bmatrix} AH_3(\beta) & BH_3(\beta) & CH_3(\beta) \\ CH_3(\beta) & AH_3(\beta) & BH_3(\beta) \\ BH_3(\beta) & CH_3(\beta) & AH_3(\beta) \end{bmatrix} \right)$$

$$+ \log m_{[e]} \left(\begin{bmatrix} AH_3(\chi) & CH_3(\chi) & BH_3(\chi) \\ BH_3(\chi) & AH_3(\chi) & CH_3(\chi) \\ CH_3(\chi) & BH_3(\chi) & AH_3(\chi) \end{bmatrix} \right)$$

$$= \log m_{[e]} \left(\begin{bmatrix} h & 0 & 0 \\ 0 & h & 0 \\ 0 & 0 & h \end{bmatrix} \right) + \log m_{[e]} \left(\exp \left(\begin{bmatrix} 0 & \beta & 0 \\ 0 & 0 & \beta \\ \beta & 0 & 0 \end{bmatrix} \right) \right) + \log m_{[e]} \left(\exp \left(\begin{bmatrix} 0 & 0 & \chi \\ \chi & 0 & 0 \\ 0 & \chi & 0 \end{bmatrix} \right) \right)$$

$$= \log m_{[e]} \left(\begin{bmatrix} h & 0 & 0 \\ 0 & h & 0 \\ 0 & 0 & h \end{bmatrix} \right) + \begin{bmatrix} 0 & \beta & 0 \\ 0 & 0 & \beta \\ \beta & 0 & 0 \end{bmatrix} + \begin{bmatrix} 0 & 0 & \chi \\ \chi & 0 & 0 \\ 0 & \chi & 0 \end{bmatrix}$$

$$= \begin{bmatrix} \log_e h & \beta & \chi \\ \chi & \log_e h & \beta \\ \beta & \chi & \log_e h \end{bmatrix}$$

In non-matrix notation, this is:

$$\log_e \left(a + b\hat{p} + c\hat{q} \right) = \log_e h + \beta\hat{p} + \chi\hat{q}$$

Section 7 : Chapter 10

3-dimensional Vectors

We take two vectors from the $C_3 L^1 H^2{}_{(j=1,k=1)}$ algebra:

$$\left\{ \begin{bmatrix} a & b & c \\ c & a & b \\ b & c & a \end{bmatrix}, \begin{bmatrix} d & e & f \\ f & d & e \\ e & f & d \end{bmatrix} \right\}$$

We normalise them:

$$\left\{ \frac{1}{\sqrt[3]{a^3+b^3+c^3-3abc}} \begin{bmatrix} a & b & c \\ c & a & b \\ b & c & a \end{bmatrix}, \frac{1}{\sqrt[3]{d^3+e^3+f^3-3def}} \begin{bmatrix} d & e & f \\ f & d & e \\ e & f & d \end{bmatrix} \right\}$$

We take the conjugate of one (either) of the normalised matrices:

$$\begin{bmatrix} \dfrac{a^2-bc}{\left(\sqrt[3]{a^3+b^3+c^3-3abc}\right)^2} & \dfrac{ab-c^2}{\left(\sqrt[3]{a^3+b^3+c^3-3abc}\right)^2} & \dfrac{ac-b^2}{\left(\sqrt[3]{a^3+b^3+c^3-3abc}\right)^2} \\[2em] \dfrac{ac-b^2}{\left(\sqrt[3]{a^3+b^3+c^3-3abc}\right)^2} & \dfrac{a^2-bc}{\left(\sqrt[3]{a^3+b^3+c^3-3abc}\right)^2} & \dfrac{ab-c^2}{\left(\sqrt[3]{a^3+b^3+c^3-3abc}\right)^2} \\[2em] \dfrac{ab-c^2}{\left(\sqrt[3]{a^3+b^3+c^3-3abc}\right)^2} & \dfrac{ac-b^2}{\left(\sqrt[3]{a^3+b^3+c^3-3abc}\right)^2} & \dfrac{a^2-bc}{\left(\sqrt[3]{a^3+b^3+c^3-3abc}\right)^2} \end{bmatrix}$$

We note that the normalisation factor is squared. We multiply this conjugate by the other normalised matrix to get:

$$\begin{bmatrix} \dfrac{a^2d+b^2e+c^2f-abf-ace-bcd}{\left(\sqrt[3]{a^3+b^3+c^3-3abc}\right)^2 \sqrt[3]{d^3+e^3+f^3-3def}} & \sim & \sim \\[2em] \dfrac{a^2f+b^2d+c^2e-abe-acd-bcf}{\left(\sqrt[3]{a^3+b^3+c^3-3abc}\right)^2 \sqrt[3]{d^3+e^3+f^3-3def}} & \sim & \sim \\[2em] \dfrac{a^2e+b^2f+c^2d-abd-acf-bce}{\left(\sqrt[3]{a^3+b^3+c^3-3abc}\right)^2 \sqrt[3]{d^3+e^3+f^3-3def}} & \sim & \sim \end{bmatrix}$$

Leading to:

$$v_{[\,]}A(\chi,\theta) = \frac{a^2d + b^2e + c^2f - abf - ace - bcd}{\left(\sqrt[3]{a^3 + b^3 + c^3 - 3abc}\right)^2 \sqrt[3]{d^3 + e^3 + f^3 - 3def}}$$

$$v_{[\,]}B(\chi,\theta) = \frac{a^2e + b^2f + c^2d - abd - acf - bce}{\left(\sqrt[3]{a^3 + b^3 + c^3 - 3abc}\right)^2 \sqrt[3]{d^3 + e^3 + f^3 - 3def}}$$

$$v_{[\,]}C(\chi,\theta) = \frac{a^2f + b^2d + c^2e - abe - acd - bcf}{\left(\sqrt[3]{a^3 + b^3 + c^3 - 3abc}\right)^2 \sqrt[3]{d^3 + e^3 + f^3 - 3def}}$$

In the case that the two vectors are the same vector:

$$v_{[\,]}A(\chi,\theta) = 1$$
$$v_{[\,]}B(\chi,\theta) = 0$$
$$v_{[\,]}C(\chi,\theta) = 0$$

In this case, $\theta = \chi = 0$, of course.

We refer to the expression:

$$a^2d + b^2e + c^2f - abf - ace - bcd$$

as the inner product. Calculating the inner product of a vector with itself leads to the distance function of the space. We refer to the expressions:

$$a^2e + b^2f + c^2d - abd - acf - bce$$
$$a^2f + b^2d + c^2e - abe - acd - bcf$$

as the exterior products. In the polar form of this algebra, the $\{b,c\}$ sub-matrices are the rotation matrices. The exterior products that occur in the $\{b,c\}$ positions in the angle product matrix are concerned with rotation in the corresponding planes analogously to the 2-dimensional cases.

If we had taken the conjugate of the other matrix, we would have come to:

$$\begin{bmatrix} \dfrac{ad^2 + be^2 + cf^2 - aef - cde - bdf}{\left(\sqrt[3]{a^3 + b^3 + c^3 - 3abc}\right)\left(\sqrt[3]{d^3 + e^3 + f^3 - 3def}\right)^2} & \sim & \sim \\[4ex] \dfrac{ae^2 + bf^2 + cd^2 - adf - bde - cef}{\left(\sqrt[3]{a^3 + b^3 + c^3 - 3abc}\right)\left(\sqrt[3]{d^3 + e^3 + f^3 - 3def}\right)^2} & \sim & \sim \\[4ex] \dfrac{af^2 + bd^2 + ce^2 - ade - bef - cdf}{\left(\sqrt[3]{a^3 + b^3 + c^3 - 3abc}\right)\left(\sqrt[3]{d^3 + e^3 + f^3 - 3def}\right)^2} & \sim & \sim \end{bmatrix}$$

which is the conjugate of the matrix we first came to.

Section 7 : Chapter 11

Conjugation in 3-dimensions

We have that the conjugate of an algebraic matrix form is the adjoint matrix. Within the 2-dimensional natural algebras, we have that multiplying a vector by its conjugate gives the length squared of that vector:

$$A * Conj(A) = length^2(A)$$

Within the 2-dimensional natural algebras, we also have that the conjugate of the conjugate of a vector is the vector:

$$Conj(Conj(A)) = A$$

Within the 3-dimensional $C_3 L^1 H^2_{(j=1,k=1)}$ algebra, we have that multiplying a vector by its conjugate gives the length cubed of that vector analogously to the 2-dimensional case:

$$B * Conj(B) = length^3(B)$$

However, the conjugate of the conjugate of a vector is not (generally) the vector. Instead, it is the vector multiplied by the length of that vector cubed:

$$Conj(Conj(B)) = B * length^3(B)$$

Within the 2-dimensional natural algebras, we have:

$$Conj(Conj(Conj(A))) = Conj(A)$$
$$Conj(Conj(Conj(Conj(A)))) = A$$

Within the 3-dimensional $C_3 L^1 H^2_{(j=1,k=1)}$ algebra, we have:

$$Conj(Conj(Conj(B))) = length^6(B) * Conj(B)$$
$$Conj(Conj(Conj(Conj(B)))) = length^{15}(B) * B$$

The same results apply to the other 3-dimensional natural algebras.

Introduction to Section 8

In this section, we examine the 3-dimensional trigonometric functions more deeply.

Section 8 : Chapter 1

The 3-dimensional Simple Trigonometric Functions

We have previously met the 3-dimensional simple trigonometric functions in the simple polar forms of the 3-dimensional flat algebras. We think of the 3-dimensional simple trigonometric functions as being projection functions that project from the unit sphere on to a 2-dimensional plane formed by two axes.

We defined the 3-dimensional simple trigonometric functions as 3-ways splittings of the exponential series (with some minus signs chucked in for the E-type). Those definitions are:

$$AH_3(x) = hypergeom\left([\ ],\left[\frac{1}{3},\frac{2}{3}\right],\left(\frac{x}{3}\right)^3\right) = \frac{x^0}{0!} + \frac{x^3}{3!} + \frac{x^6}{6!} + \frac{x^9}{9!} + ...$$

$$BH_3(x) = \frac{x}{1!}\left(hypergeom\left([\ ],\left[\frac{2}{3},\frac{4}{3}\right],\left(\frac{x}{3}\right)^3\right)\right) = \frac{x^1}{1!} + \frac{x^4}{4!} + \frac{x^7}{7!} + \frac{x^{10}}{10!} + ...$$

$$CH_3(x) = \frac{x^2}{2!}\left(hypergeom\left([\ ],\left[\frac{4}{3},\frac{5}{3}\right],\left(\frac{x}{3}\right)^3\right)\right) = \frac{x^2}{2!} + \frac{x^5}{5!} + \frac{x^8}{8!} + \frac{x^{11}}{11!} + ...$$

$$AE_3(x) = hypergeom\left([\ ],\left[\frac{1}{3},\frac{2}{3}\right],-\left(\frac{x}{3}\right)^3\right) = \frac{x^0}{0!} - \frac{x^3}{3!} + \frac{x^6}{6!} - \frac{x^9}{9!} + ...$$

$$BE_3(x) = \frac{x}{1!}\left(hypergeom\left([\ ],\left[\frac{2}{3},\frac{4}{3}\right],-\left(\frac{x}{3}\right)^3\right)\right) = \frac{x^1}{1!} - \frac{x^4}{4!} + \frac{x^7}{7!} - \frac{x^{10}}{10!} + ...$$

$$CE_3(x) = \frac{x^2}{2!}\left(hypergeom\left([\ ],\left[\frac{4}{3},\frac{5}{3}\right],-\left(\frac{x}{3}\right)^3\right)\right) = \frac{x^2}{2!} - \frac{x^5}{5!} + \frac{x^8}{8!} - \frac{x^{11}}{11!} + ...$$

The first letter of the appellations chosen is just a letter. The second letter is either an H to denote H-type or an E to denote E-type. The sub-script number is the splitting of the exponential; it is also the dimension of the natural space of the algebra. Using the above nomenclature, the familiar 2-dimensional trigonometric functions would be:

$$\{\cos \equiv AE_2,\ \sin \equiv BE_2,\ \cosh \equiv AH_2,\ \sinh \equiv BH_2\}$$

and the 1-dimensional trigonometric function, 1, might be written as AH_1.

We immediately have:

$$e^x = AH_3(x) + BH_3(x) + CH_3(x)$$
$$e^{-x} = AE_3(x) - BE_3(x) + CE_3(x)$$

Hence:

$$\left(AH_3(x) + BH_3(x) + CH_3(x)\right)\left(AE_3(x) - BE_3(x) + CE_3(x)\right) = 1$$

Other identities are:

$$AE_3(x)BE_3(x) - AE_3(x)CE_3(x) + BE_3(x)CE_3(x) = \frac{1}{3}\left(-e^{-2x} + e^x\right)$$

$$AH_3(x)BH_3(x) + AH_3(x)CH_3(x) + BH_3(x)CH_3(x) = \frac{1}{3}\left(e^{2x} - e^{-x}\right)$$

We also have:

$$AH_3(-x) = AE_3(x)$$
$$BH_3(-x) = -BE_3(x)$$
$$CH_3(-x) = CE_3(x)$$
$$AE_3(-x) = AH_3(x)$$
$$BE_3(-x) = -BH_3(x)$$
$$CE_3(-x) = CH_3(x)$$

Differentiation:

Clearly[64], we have cyclic differential relations similar to *sinh* & *cosh* and *sin* & *cos* but with a 6 & 3-cycle rather than a 4 & 2-cycle.

$$\left\{\begin{aligned}
&\frac{d}{dx}CE_3x = BE_3x, && \frac{d^2}{dx^2}CE_3x = AE_3x, && \frac{d^3}{dx^3}CE_3x = -CE_3x, \\
&\frac{d^4}{dx^4}CE_3x = -BE_3x, && \frac{d^5}{dx^5}CE_3x = -AE_3x, && \frac{d^6}{dx^6}CE_3x = CE_3x
\end{aligned}\right\}$$

$$\left\{\frac{d}{dx}CH_3x = BH_3x, \quad \frac{d^2}{dx^2}CH_3x = AH_3x, \quad \frac{d^3}{dx^3}CH_3x = CH_3x\right\}$$

In Standard Function Form:

These simple trigonometric functions have representation as standard functions:

$$AH_3(x) = \frac{1}{3}\left(e^x + 2e^{-\left(\frac{x}{2}\right)}\cos\left(\frac{\sqrt{3}}{2}x\right)\right)$$

$$BH_3(x) = \frac{1}{3}\left(e^x - 2e^{-\left(\frac{x}{2}\right)}\cos\left(\frac{\sqrt{3}}{2}x + \frac{\pi}{3}\right)\right)$$

$$CH_3(x) = \frac{1}{3}\left(e^x + 2e^{-\left(\frac{x}{2}\right)}\cos\left(\frac{\sqrt{3}}{2}x + \frac{2\pi}{3}\right)\right)$$

---

[64] It is clear when you think of the series representation.

117

$$AE_3(x) = \frac{1}{3}\left(e^{-x} + 2e^{\left(\frac{x}{2}\right)} \cos\left(\frac{\sqrt{3}}{2}x \right) \right)$$

$$BE_3(x) = -\frac{1}{3}\left(e^{-x} + 2e^{\left(\frac{x}{2}\right)} \cos\left(\frac{\sqrt{3}}{2}x + \frac{2\pi}{3} \right) \right)$$

$$CE_3(x) = \frac{1}{3}\left(e^{-x} - 2e^{\left(\frac{x}{2}\right)} \cos\left(\frac{\sqrt{3}}{2}x + \frac{\pi}{3} \right) \right)$$

Graphs: H-Type
When $x = 0$:

$$AH_3(0) = 1$$
$$BH_3(0) = 0$$
$$CH_3(0) = 0$$

AH_3 has a point of inflection at $x = 0$. CH_3 has a local minimum at $x = 0$. BH_3 passes through 0 with slope 1. These functions are wave-like for small values of x. They are wave-like with enormous amplitude for large negative x, but, for large positive x, the wave-like nature is swamped and they appear non-wavelike unless examined minutely.

The wavy parts of the 3-dimensional simple trigonometric functions are out of phase with each other by $\frac{\pi}{3}$. However, the presence of the exponential parts of these functions means that they cannot be translated along the x-axis to sit on top of each other in the way the 2-dimensional euclidean trigonometric functions can be.

Graph of AH3, BH3, & CH3

Graph of AH3, BH3, & CH3

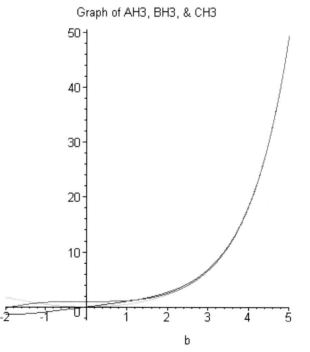

Graphs: E-type
When $x = 0$:

$$AE_3(0) = 1$$
$$BE_3(0) = 0$$
$$CE_3(0) = 0$$

AE_3 has a point of inflection at $x = 0$. CE_3 has a local minimum at $x = 0$. BE_3 passes through 0 with slope 1. These functions are wave-like with enormous amplitude for large positive values of x. For large negative x, the wave-like nature is swamped and they appear non-wavelike unless examined minutely. BE_3 grows exponentially negative while AE_3 and CE_3 grow exponentially positive.

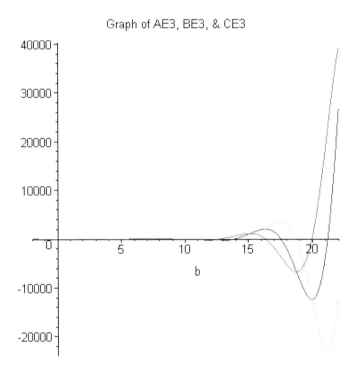

Graph of AE3, BE3, & CE3

Graph of AE3, BE3, & CE3

It is left as an exercise for the reader to do Fourier analysis with these simple trigonometric functions.

Section 8 : Chapter 2

The $C_3 L^1 H^2_{(j=1,k=1)}$ v-Functions

We have met the 3-dimensional v-functions previously. Each of the four 3-dimensional flat algebras has a set of v-functions. Although it is possible to construct a general 3-dimensional v-function from the general 3-dimensional polar form, the expression is very complicated and we choose not to include it in this work.

The 3-dimensional v-functions are projections from the unit sphere on to the axes of the space.

The $C_3 L^1 H^2_{(j=1,k=1)}$ v-functions are:

$$v_{\left[C_3 L^1 H^2_{(j=1,k=1)}\right]} A(b,c) = \frac{1}{3}\left(e^{(b+c)} + 2e^{-\left(\frac{b+c}{2}\right)} \cos\left(\frac{\sqrt{3}}{2}(b-c) \right) \right)$$

$$v_{\left[C_3 L^1 H^2_{(j=1,k=1)}\right]} B(b,c) = \frac{1}{3}\left(e^{(b+c)} - 2e^{-\left(\frac{b+c}{2}\right)} \cos\left(\frac{\sqrt{3}}{2}(b-c) + \frac{\pi}{3} \right) \right)$$

$$v_{\left[C_3 L^1 H^2_{(j=1,k=1)}\right]} C(b,c) = \frac{1}{3}\left(e^{(b+c)} + 2e^{-\left(\frac{b+c}{2}\right)} \cos\left(\frac{\sqrt{3}}{2}(b-c) + \frac{2\pi}{3} \right) \right)$$

When we set one of the variables to zero, we reduce the v-functions to the simple trigonometric functions. We have:

$$v_{\left[C_3 L^1 H^2_{(j=1,k=1)}\right]} A(b,c) + v_{\left[C_3 L^1 H^2_{(j=1,k=1)}\right]} B(b,c) + v_{\left[C_3 L^1 H^2_{(j=1,k=1)}\right]} C(b,c) = e^{(b+c)}$$

but the v-functions are not simple splittings of the exponential series of e^{b+c}. It is the simple trigonometric functions that are the simple splittings of the exponential series.

Differentiation:
We have:

$$\frac{\partial\left(v_{\left[C_3 L^1 H^2_{(j=1,k=1)}\right]} A(b,c) \right)}{\partial b} = v_{\left[C_3 L^1 H^2_{(j=1,k=1)}\right]} C(b,c)$$

$$\frac{\partial\left(v_{\left[C_3 L^1 H^2_{(j=1,k=1)}\right]} A(b,c) \right)}{\partial c} = v_{\left[C_3 L^1 H^2_{(j=1,k=1)}\right]} B(b,c)$$

$$\frac{\partial\left(v_{\left[C_3 L^1 H^2_{(j=1,k=1)}\right]} B(b,c) \right)}{\partial b} = v_{\left[C_3 L^1 H^2_{(j=1,k=1)}\right]} A(b,c)$$

$$\frac{\partial\left(v_{\left[C_3 L^1 H^2_{(j=1,k=1)}\right]} B(b,c) \right)}{\partial c} = v_{\left[C_3 L^1 H^2_{(j=1,k=1)}\right]} C(b,c)$$

$$\frac{\partial\left(v_{\left[C_3 L^l H^2_{(j=1,k=1)}\right]}C(b,c)\right)}{\partial b} = v_{\left[C_3 L^l H^2_{(j=1,k=1)}\right]}B(b,c)$$

$$\frac{\partial\left(v_{\left[C_3 L^l H^2_{(j=1,k=1)}\right]}C(b,c)\right)}{\partial c} = v_{\left[C_3 L^l H^2_{(j=1,k=1)}\right]}A(b,c)$$

Thus we have a differentiation cycle.

Anything with a repeating differentiation cycle just has[65] to be something to do with the exponential series. We have:

Expanding about $c = 0$:

$$v_{\left[C_3 L^l H^2_{(j=1,k=1)}\right]}A(b,c) = \frac{[AH_3(b)]c^0}{0!} + \frac{[BH_3(b)]c^1}{1!} + \frac{[CH_3(b)]c^2}{2!} + \frac{[AH_3(b)]c^3}{3!} + ...$$

$$= \sum_{n=0}^{n=\infty}\left(\frac{[AH_3(b)]c^{3n}}{(3n)!} + \frac{[BH_3(b)]c^{3n+1}}{(3n+1)!} + \frac{[CH_3(b)]c^{3n+2}}{(3n+2)!}\right)$$

Inspection of this reveals that differentiation with respect to either variable (3 times) leads to the same series, of course. Similar series arise from expanding about the other variables and for the other v-functions.

Graphs:
Small scale:

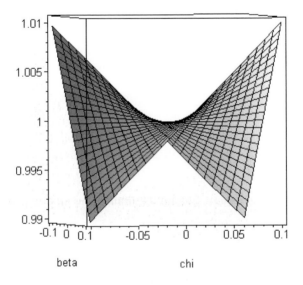

beta   chi

---

[65] This is an emotional outburst not a mathematical fact.

Large scale:

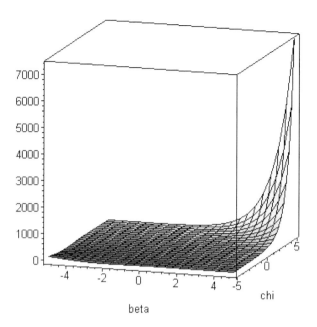

All three functions plotted in the same axes are a tsunami graph for small and negative values of *b* and *c*, but, for large positive values of *b* and *c*, the functions join together to race off exponentially toward infinity.

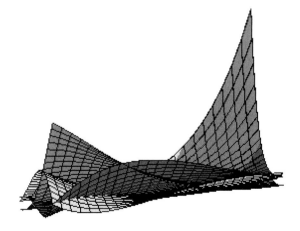

Similar results apply to the other 3-dimensional algebras.

Introduction to Section 9

In this section, we meet the 4-dimensional natural algebras and the 4-dimensional simple trigonometric functions. Much within this section is analogous to the 3-dimensional natural algebras, but there are some differences. There is only one basic 3-dimensional algebraic matrix form. There are three plus one basic algebraic matrix forms in four dimensions. The 3-dimensional natural algebras contain only the real numbers as a sub-algebra. The 4-dimensional natural algebras contain, as well as the real numbers, 2-dimensional sub-algebras.

Section 9 : Chapter 1

The Polar Form of the $C_{4.1}L^lH^3{}_{(j=1,k=1,l=1)}$ Algebra

The general form of the $C_{4.1}$ algebra is:

$$C_{4.1} = \begin{bmatrix} a & b & c & d \\ jd & a & kb & lc \\ \dfrac{jl}{k}c & \dfrac{j}{k}d & a & lb \\ jb & \dfrac{j}{k}c & \dfrac{j}{l}d & a \end{bmatrix}$$

We take $\{ j=1, k=1, l=1 \}$ and exponentiate this matrix.

$$\exp\left(\begin{bmatrix} a & b & c & d \\ d & a & b & c \\ c & d & a & b \\ b & c & d & a \end{bmatrix}\right) = PROD \left(\begin{bmatrix} e^a & 0 & 0 & 0 \\ 0 & e^a & 0 & 0 \\ 0 & 0 & e^a & 0 \\ 0 & 0 & 0 & e^a \end{bmatrix} \begin{bmatrix} AH_4(b) & BH_4(b) & CH_4(b) & DH_4(b) \\ DH_4(b) & AH_4(b) & BH_4(b) & CH_4(b) \\ CH_4(b) & DH_4(b) & AH_4(b) & BH_4(b) \\ BH_4(b) & CH_4(b) & DH_4(b) & AH_4(b) \end{bmatrix} \right.$$

$$\begin{bmatrix} \cosh(c) & 0 & \sinh(c) & 0 \\ 0 & \cosh(c) & 0 & \sinh(c) \\ \sinh(c) & 0 & \cosh(c) & 0 \\ 0 & \sinh(c) & 0 & \cosh(c) \end{bmatrix}$$

$$\left. \begin{bmatrix} AH_4(d) & DH_4(d) & CH_4(d) & BH_4(d) \\ BH_4(d) & AH_4(d) & DH_4(d) & CH_4(d) \\ CH_4(d) & BH_4(d) & AH_4(d) & DH_4(d) \\ DH_4(d) & CH_4(d) & BH_4(d) & AH_4(d) \end{bmatrix} \right)$$

Where:

$$AH_4(x) = 1 + \frac{x^4}{4!} + \frac{x^8}{8!} + \ldots$$

$$BH_4(x) = x + \frac{x^5}{5!} + \frac{x^9}{9!} + \ldots$$

$$CH_4(x) = \frac{x^2}{2!} + \frac{x^6}{6!} + \frac{x^{10}}{10!} + \ldots$$

$$DH_4(x) = \frac{x^3}{3!} + \frac{x^7}{7!} + \frac{x^{11}}{11!} + \ldots$$

are the 4-way splittings of the exponential series that are the 4-dimensional simple trigonometric functions.

The matrices containing the simple trigonometric functions were formed by exponentiating a matrix with trace zero and thus the determinants of these matrices are unity. Thereby do we have a (large) trigonometric identity.

$$AH_4^4(x) - BH_4^4(x) + CH_4^4(x) - DH_4^4(x) - 2AH_4^2(x)CH_4^2(x)$$
$$+2BH_4^2(x)DH_4^2(x) - 4AH_4^2(x)BH_4(x)DH_4(x)$$
$$+4AH_4(x)BH_4^2(x)CH_4(x) - 4BH_4(x)CH_4^2(x)DH_4(x)$$
$$+4AH_4(x)CH_4(x)DH_4^2(x) = 1$$

We also have:

$$\begin{bmatrix} AH_4(b+x) & BH_4(b+x) & CH_4(b+x) & DH_4(b+x) \\ DH_4(b+x) & AH_4(b+x) & BH_4(b+x) & CH_4(b+x) \\ CH_4(b+x) & DH_4(b+x) & AH_4(b+x) & BH_4(b+x) \\ BH_4(b+x) & CH_4(b+x) & DH_4(b+x) & AH_4(b+x) \end{bmatrix}$$

$$= PROD \left(\begin{bmatrix} AH_4(b) & BH_4(b) & CH_4(b) & DH_4(b) \\ DH_4(b) & AH_4(b) & BH_4(b) & CH_4(b) \\ CH_4(b) & DH_4(b) & AH_4(b) & BH_4(b) \\ BH_4(b) & CH_4(b) & DH_4(b) & AH_4(b) \end{bmatrix} \begin{bmatrix} AH_4(x) & BH_4(x) & CH_4(x) & DH_4(x) \\ DH_4(x) & AH_4(x) & BH_4(x) & CH_4(x) \\ CH_4(x) & DH_4(x) & AH_4(x) & BH_4(x) \\ BH_4(x) & CH_4(x) & DH_4(x) & AH_4(x) \end{bmatrix} \right) \quad [66]$$

Leading to the trigonometric identities:

---

[66] For calculation details, readers are referred to the chapter on the polar form of the $C_3 L^1 H^2_{(j=1,k=1)}$ algebra.

$$AH_4(b+x) =$$
$$AH_4(b)AH_4(x) + BH_4(b)DH_4(x) + DH_4(b)BH_4(x) + CH_4(b)CH_4(x)$$

$$BH_4(b+x) =$$
$$AH_4(b)BH_4(x) + BH_4(b)AH_4(x) + CH_4(b)DH_4(x) + DH_4(b)CH_4(x)$$

$$CH_4(b+x) =$$
$$AH_4(b)CH_4(x) + BH_4(b)BH_4(x) + CH_4(b)AH_4(x) + DH_4(b)DH_4(x)$$

$$DH_4(b+x) =$$
$$AH_4(b)DH_4(x) + BH_4(b)CH_4(x) + CH_4(b)BH_4(x) + DH_4(b)AH_4(x)$$

We have:

$$
\det\left(\begin{bmatrix} w & x & y & z \\ z & w & x & y \\ y & z & w & x \\ x & y & z & w \end{bmatrix}\right) = \det\left(PROD\left(\begin{bmatrix} r & 0 & 0 & 0 \\ 0 & r & 0 & 0 \\ 0 & 0 & r & 0 \\ 0 & 0 & 0 & r \end{bmatrix}, \begin{bmatrix} AH_4(b) & BH_4(b) & CH_4(b) & DH_4(b) \\ DH_4(b) & AH_4(b) & BH_4(b) & CH_4(b) \\ CH_4(b) & DH_4(b) & AH_4(b) & BH_4(b) \\ BH_4(b) & CH_4(b) & DH_4(b) & AH_4(b) \end{bmatrix}, \begin{bmatrix} \cosh(c) & 0 & \sinh(c) & 0 \\ 0 & \cosh(c) & 0 & \sinh(c) \\ \sinh(c) & 0 & \cosh(c) & 0 \\ 0 & \sinh(c) & 0 & \cosh(c) \end{bmatrix}, \begin{bmatrix} AH_4(d) & DH_4(d) & CH_4(d) & BH_4(d) \\ BH_4(d) & AH_4(d) & DH_4(d) & CH_4(d) \\ CH_4(d) & BH_4(d) & AH_4(d) & DH_4(d) \\ DH_4(d) & CH_4(d) & BH_4(d) & AH_4(d) \end{bmatrix}\right)\right)
$$

Leading to the distance function:

$$r = \sqrt[4]{a^4 - b^4 + c^4 - d^4 - 2a^2c^2 + 2b^2d^2 - 4a^2bd + 4ab^2c - 4bc^2d + 4acd^2}$$

When summed, the 4-dimensional simple trigonometric functions have representation:

$$AH_4(x) = \frac{1}{4}\left(2\cosh(x) + 2\cos(x)\right)$$

$$BH_4(x) = \frac{1}{4}\left(2\sinh(x) + 2\sin(x)\right)$$

$$CH_4(x) = \frac{1}{4}\left(2\cosh(x) - 2\cos(x)\right)$$

$$DH_4(x) = \frac{1}{4}\left(2\sinh(x) - 2\sin(x)\right)$$

The reader might be surprised by the occurrence of the 2-dimensional hyperbolic trigonometric functions within a matrix in the polar form of this 4-dimensional algebra. Such a matrix occurs in the polar form of every $\{C_{4.1}, C_{4.2}, C_{4.3}\}$ algebra. The polar form of the $C_2 \times C_2$ algebras has the form of three such matrices containing 2-dimensional trigonometric functions. The presence of these matrices is directly associated with the fact that the $\{C_{4.1}, C_{4.2}, C_{4.3}\}$ algebras have one 2-dimensional sub-algebra and the $C_2 \times C_2$ algebra has three 2-dimensional sub-algebras.

Section 9 : Chapter 2

The Polar Form of the $C_{4.1}L^1E^3_{(j=-1,k=1,l=1)}$ Algebra

We have:

$$\exp\left(\begin{bmatrix} a & b & c & d \\ -d & a & b & c \\ -c & -d & a & b \\ -b & -c & -d & a \end{bmatrix}\right) = PROD \left(\begin{bmatrix} e^a & 0 & 0 & 0 \\ 0 & e^a & 0 & 0 \\ 0 & 0 & e^a & 0 \\ 0 & 0 & 0 & e^a \end{bmatrix} \begin{bmatrix} AE_4(b) & BE_4(b) & CE_4(b) & DE_4(b) \\ -DE_4(b) & AE_4(b) & BE_4(b) & CE_4(b) \\ -CE_4(b) & -DE_4(b) & AE_4(b) & BE_4(b) \\ -BE_4(b) & -CE_4(b) & -DE_4(b) & AE_4(b) \end{bmatrix} \begin{bmatrix} \cos(c) & 0 & \sin(c) & 0 \\ 0 & \cos(c) & 0 & \sin(c) \\ -\sin(c) & 0 & \cos(c) & 0 \\ 0 & -\sin(c) & 0 & \cos(c) \end{bmatrix} \begin{bmatrix} AE_4(d) & DE_4(d) & -CE_4(d) & BE_4(d) \\ -BE_4(d) & AE_4(d) & DE_4(d) & -CE_4(d) \\ CE_4(d) & -BE_4(d) & AE_4(d) & DE_4(d) \\ -DE_4(d) & CE_4(d) & -BE_4(d) & AE_4(d) \end{bmatrix} \right)$$

Where:

$$AE_4(x) = 1 - \frac{x^4}{4!} + \frac{x^8}{8!} - \dots$$

$$BE_4(x) = x - \frac{x^5}{5!} + \frac{x^9}{9!} - \dots$$

$$CE_4(x) = \frac{x^2}{2!} - \frac{x^6}{6!} + \frac{x^{10}}{10!} - \dots$$

$$DE_4(x) = \frac{x^3}{3!} - \frac{x^7}{7!} + \frac{x^{11}}{11!} - \dots$$

are the 4-way splittings of the exponential series (with some minus signs chucked in) that are the 4-dimensional E-type simple trigonometric functions.

We have the customary trigonometric identities

$$AE_4^{\;4}(x)+BE_4^{\;4}(x)+CE_4^{\;4}(x)+DE_4^{\;4}(x)+2AE_4^{\;2}(x)CE_4^{\;2}(x)+2BE_4^{\;2}(x)DE_4^{\;2}(x)$$
$$+4AE_4^{\;2}(x)BE_4(x)DE_4(x)-4AE_4(x)BE_4^{\;2}(x)CE_4(x)-4BE_4(x)CE_4^{\;2}(x)DE_4(x)$$
$$+4AE_4(x)CE_4(x)DE_4^{\;2}(x)=1$$

$$AE_4(b+x)=AE_4(b)AE_4(x)-BE_4(b)DE_4(x)-DE_4(b)BE_4(x)-CE_4(b)CE_4(x)$$
$$BE_4(b+x)=AE_4(b)BE_4(x)+BE_4(b)AE_4(x)-CE_4(b)DE_4(x)-DE_4(b)CE_4(x)$$
$$CE_4(b+x)=AE_4(b)CE_4(x)+BE_4(b)BE_4(x)+CE_4(b)AE_4(x)-DE_4(b)DE_4(x)$$
$$DE_4(b+x)=AE_4(b)DE_4(x)+BE_4(b)CE_4(x)+CE_4(b)BE_4(x)+DE_4(b)AE_4(x)$$

We have the distance function:

$$r=\sqrt[4]{a^4+b^4+c^4+d^4+2a^2c^2+2b^2d^2+4a^2bd-4ab^2c-4bc^2d+4acd^2}$$

When summed, the 4-dimensional E-Type simple trigonometric functions have representation:

$$AE_4(x)=\cos\left(\frac{x}{\sqrt{2}}\right)\cosh\left(\frac{x}{\sqrt{2}}\right)$$

$$BE_4(x)=\frac{1}{\sqrt{2}}\left(\cos\left(\frac{x}{\sqrt{2}}\right)\sinh\left(\frac{x}{\sqrt{2}}\right)+\sin\left(\frac{x}{\sqrt{2}}\right)\cosh\left(\frac{x}{\sqrt{2}}\right)\right)$$

$$CE_4(x)=\sin\left(\frac{x}{\sqrt{2}}\right)\sinh\left(\frac{x}{\sqrt{2}}\right)$$

$$DE_4(x)=\frac{1}{\sqrt{2}}\left(-\cos\left(\frac{x}{\sqrt{2}}\right)\sinh\left(\frac{x}{\sqrt{2}}\right)+\sin\left(\frac{x}{\sqrt{2}}\right)\cosh\left(\frac{x}{\sqrt{2}}\right)\right)$$

Section 9 : Chapter 3

The Polar Forms of the $\{C_{4.1}, C_{4.2}, C_{4.3}\}$ Algebras in General

We have previously shown that the

$$\left\{C_{4.1}L^{l}H^{3}{}_{(j=1,k=1,l=1)}, C_{4.2}L^{l}H^{3}{}_{(j=1,k=1,l=1)}, C_{4.3}L^{l}H^{3}{}_{(j=1,k=1,l=1)}\right\}$$

algebras are algebraically isomorphic to each other[67]. The reader is directed to inspect the appendices 5.1, 5.2, & 5.3 listing the polar forms of the flat $\{C_{4.1}, C_{4.2}, C_{4.3}\}$ algebras. The eight flat polar forms of the $C_{4.1}$ algebra all have the same basic form. Other than the type (E-type or H-type) of simple trigonometric functions in the simple polar form matrices, they differ only in the way that the minus signs are scattered about. The same is true of the $\{C_{4.2}, C_{4.3}\}$ algebras.

In two of the simple polar forms of the 3-dimensional flat algebras, we found both H-type and E-type simple trigonometric functions (though not in the same matrix). In the 4-dimensional natural algebras $\{C_{4.1}, C_{4.2}, C_{4.3}\}$, such mixing never occurs.[68]

The occurrence of a matrix containing the 2-dimensional trigonometric functions is connected to the existence of a 2-dimensional sub-algebra within the $\{C_{4.1}, C_{4.2}, C_{4.3}\}$ algebras. In the case of the $C_{4.1}$ algebras, this sub-algebra is formed from the $\{a,c\}$ variables.

$$\begin{bmatrix} a & 0 & c & 0 \\ 0 & a & 0 & lc \\ \dfrac{jl}{k}c & 0 & a & 0 \\ 0 & \dfrac{j}{k}c & 0 & a \end{bmatrix} \begin{bmatrix} e & 0 & f & 0 \\ 0 & e & 0 & lf \\ \dfrac{jl}{k}f & 0 & e & 0 \\ 0 & \dfrac{j}{k}f & 0 & e \end{bmatrix} =$$

$$\begin{bmatrix} ae + \dfrac{jl}{k}cf & 0 & af + ce & 0 \\ 0 & ae + \dfrac{jl}{k}cf & 0 & l(af + ce) \\ \dfrac{jl}{k}(af + ce) & 0 & ae + \dfrac{jl}{k}cf & 0 \\ 0 & \dfrac{j}{k}(af + ce) & 0 & ae + \dfrac{jl}{k}cf \end{bmatrix}$$

In the case of the $C_{4.2}$ algebras, this sub-algebra is formed from the $\{a,b\}$ variables, and in the case of the $C_{4.3}$ algebras, this sub-algebra is formed from the $\{a,d\}$ variables.[69]

---

[67] Section 5, Chapter 1, Addendum 2.

[68] It does occur in some of the polar forms of the $C_{2} \times C_{2}$ algebra.

Section 9 : Chapter 4

The Polar Forms of the $C_2 \times C_2$ Algebras

The reader is directed to inspect the polar forms of these algebras in appendix 5.4. We have:

$$\exp\left(\begin{bmatrix} a & b & c & d \\ -b & a & -d & c \\ -c & -d & a & b \\ d & -c & -b & a \end{bmatrix}\right) = PROD \left(\begin{bmatrix} h & 0 & 0 & 0 \\ 0 & h & 0 & 0 \\ 0 & 0 & h & 0 \\ 0 & 0 & 0 & h \end{bmatrix} \begin{bmatrix} \cos(b) & \sin(b) & 0 & 0 \\ -\sin(b) & \cos(b) & 0 & 0 \\ 0 & 0 & \cos(b) & \sin(b) \\ 0 & 0 & -\sin(b) & \cos(b) \end{bmatrix} \right.$$

$$\begin{bmatrix} \cos(c) & 0 & \sin(c) & 0 \\ 0 & \cos(c) & 0 & \sin(c) \\ -\sin(c) & 0 & \cos(c) & 0 \\ 0 & -\sin(c) & 0 & \cos(c) \end{bmatrix}$$

$$\left. \begin{bmatrix} \cosh(d) & 0 & 0 & \sinh(d) \\ 0 & \cosh(d) & -\sinh(d) & 0 \\ 0 & -\sinh(d) & \cosh(d) & 0 \\ \sinh(d) & 0 & 0 & \cosh(d) \end{bmatrix} \right)$$

This polar form has three rotation matrices containing 2-dimensional trigonometric functions. All $C_2 \times C_2$ algebras have a similar polar form with three such matrices. This contrasts with the cases of the other 4-dimensional algebraic matrix forms. In this case, as in most, there is a mixture of E-type trigonometric function matrices and H-type trigonometric function matrices.

The polar form reflects the fact that these algebras each have three 2-dimensional sub-algebras. In this case, we have:

$$\begin{bmatrix} a & b & 0 & 0 \\ -b & a & 0 & 0 \\ 0 & 0 & a & b \\ 0 & 0 & -b & a \end{bmatrix} \begin{bmatrix} e & f & 0 & 0 \\ -f & e & 0 & 0 \\ 0 & 0 & e & f \\ 0 & 0 & -f & e \end{bmatrix} = \begin{bmatrix} ae-bf & af+be & 0 & 0 \\ -(af+be) & ae-bf & 0 & 0 \\ 0 & 0 & ae-bf & af+be \\ 0 & 0 & -(af+be) & ae-bf \end{bmatrix}$$

$$\begin{bmatrix} a & 0 & c & 0 \\ 0 & a & 0 & c \\ -c & 0 & a & 0 \\ 0 & -c & 0 & a \end{bmatrix} \begin{bmatrix} e & 0 & g & 0 \\ 0 & e & 0 & g \\ -g & 0 & e & 0 \\ 0 & -g & 0 & e \end{bmatrix} = \begin{bmatrix} ae-cg & 0 & ag+ce & 0 \\ 0 & ae-cg & 0 & ag+ce \\ -(ag+ce) & 0 & ae-cg & 0 \\ 0 & -(ag+ce) & 0 & ae-cg \end{bmatrix}$$

---

[69] Each of these three algebras is essentially the same as the cyclic group C_4 with the C_2 subgroup being generated by a different one of each of the other non-identity elements.

$$\begin{bmatrix} a & 0 & 0 & d \\ 0 & a & -d & 0 \\ 0 & -d & a & 0 \\ d & 0 & 0 & a \end{bmatrix}\begin{bmatrix} e & 0 & 0 & h \\ 0 & e & -h & 0 \\ 0 & -h & e & 0 \\ h & 0 & 0 & e \end{bmatrix} = \begin{bmatrix} ae+dh & 0 & 0 & ah+de \\ 0 & ae+dh & -(ah+de) & 0 \\ 0 & -(ah+de) & ae+dh & 0 \\ ah+de & 0 & 0 & ae+dh \end{bmatrix}$$

The reader should compare these sub-algebras to the types of trigonometric functions within the polar form. There is a correspondence between the sub-algebras formed by the a variable and a particular variable and the type of trigonometric functions within the rotation matrix of that particular variable.

The distance function of this particular flat algebra is:

$$r = \sqrt[4]{\left((a-d)^2 + (b+c)^2\right)\left((a+d)^2 + (b-c)^2\right)}$$

Section 9 : Chapter 5

The General Polar Form $C_{4.1}$

We have:

$$\exp\left(\begin{bmatrix} a & b & c & d \\ jd & a & kb & lc \\ \dfrac{jl}{k}c & \dfrac{j}{k}d & a & lb \\ jb & \dfrac{j}{k}c & \dfrac{j}{l}d & a \end{bmatrix}\right) =$$

[~]

$$\begin{bmatrix}
AH_4\left((jkl)^{\frac{1}{4}}b\right) & \left(\dfrac{1}{jkl}\right)^{\frac{1}{4}}BH_4\left((jkl)^{\frac{1}{4}}b\right) & k\left(\dfrac{1}{jkl}\right)^{\frac{2}{4}}CH_4\left((jkl)^{\frac{1}{4}}b\right) & kl\left(\dfrac{1}{jkl}\right)^{\frac{3}{4}}DH_4\left((jkl)^{\frac{1}{4}}b\right) \\
jkl\left(\dfrac{1}{jkl}\right)^{\frac{3}{4}}DH_4\left((jkl)^{\frac{1}{4}}b\right) & AH_4\left((jkl)^{\frac{1}{4}}b\right) & k\left(\dfrac{1}{jkl}\right)^{\frac{1}{4}}BH_4\left((jkl)^{\frac{1}{4}}b\right) & kl\left(\dfrac{1}{jkl}\right)^{\frac{2}{4}}CH_4\left((jkl)^{\frac{1}{4}}b\right) \\
jl\left(\dfrac{1}{jkl}\right)^{\frac{2}{4}}CH_4\left((jkl)^{\frac{1}{4}}b\right) & jl\left(\dfrac{1}{jkl}\right)^{\frac{3}{4}}DH_4\left((jkl)^{\frac{1}{4}}b\right) & AH_4\left((jkl)^{\frac{1}{4}}b\right) & l\left(\dfrac{1}{jkl}\right)^{\frac{1}{4}}BH_4\left((jkl)^{\frac{1}{4}}b\right) \\
j\left(\dfrac{1}{jkl}\right)^{\frac{1}{4}}BH_4\left((jkl)^{\frac{1}{4}}b\right) & j\left(\dfrac{1}{jkl}\right)^{\frac{2}{4}}CH_4\left((jkl)^{\frac{1}{4}}b\right) & jk\left(\dfrac{1}{jkl}\right)^{\frac{3}{4}}DH_4\left((jkl)^{\frac{1}{4}}b\right) & AH_4\left((jkl)^{\frac{1}{4}}b\right)
\end{bmatrix}$$

[~]
[~]

where we have twiddled out the unwanted simple rotation matrices.

We consider the b rotation matrix. We take $\{j=-1,k=1,l=1\}$ and this is equal to the b rotation matrix of the $C_{4.1}L^lE^3{}_{(j=-1,k=1,l=1)}$ algebra. Comparing the elements within the matrices gives:

$$AH_4\left((-1)^{\frac{1}{4}}b\right)=AE_4(b)$$

$$(-1)^{\frac{1}{4}}BH_4\left((-1)^{\frac{1}{4}}b\right)=BE_4(b)$$

$$(-1)^{\frac{2}{4}}CH_4\left((-1)^{\frac{1}{4}}b\right)=CE_4(b)$$

$$(-1)^{\frac{3}{4}}DH_4\left((-1)^{\frac{1}{4}}b\right)=DE_4(b)$$

Section 9 : Chapter 6

Non-matrix Notation - $C_{4.1}L^1H^3{}_{(j=1,k=1,l=1)}$

We have:

$$C_{4.1}L^1H^3{}_{(j=1,k=1,l=1)} = \begin{bmatrix} a & b & c & d \\ d & a & b & c \\ c & d & a & b \\ b & c & d & a \end{bmatrix}$$

The non-matrix representation of this algebra is: $a+b\hat{r}+c\hat{s}+d\hat{t}$. The $\hat{r}$ is not the same object we encountered in the hyperbolic complex numbers. It is just that we are running out of letters in the alphabet. Addition is component-wise:

$$a+b\hat{r}+c\hat{s}+d\hat{t}+e+f\hat{r}+g\hat{s}+h\hat{t} = (a+e)+(b+f)\hat{r}+(c+g)\hat{s}+(d+h)\hat{t}$$

Multiplication is taken (carefully) from the matrix product:

$$\left(a+b\hat{r}+c\hat{s}+d\hat{t}\right)\left(e+f\hat{r}+g\hat{s}+h\hat{t}\right) = (ae+bh+cg+df)+(af+be+ch+dg)\hat{r}$$
$$+(ag+bf+ce+dh)\hat{s}+(ah+bg+cf+de)\hat{t}$$

The conjugate is taken from the adjoint matrix:

$$\overline{\left(a+b\hat{r}+c\hat{s}+d\hat{t}\right)} = \left(a^3+cb^2-ac^2+cd^2-2abd\right)+\left(-d^3-a^2b+b^2d-bc^2+2acd\right)\hat{r}$$
$$+\left(c^3-a^2c+ab^2+ad^2-2bcd\right)\hat{s}+\left(-b^3-a^2d-c^2d+bd^2+2abc\right)\hat{t}$$

Term by term expansion of the multiplication operation gives:

$$\left(a+b\hat{r}+c\hat{s}+d\hat{t}\right)\left(e+f\hat{r}+g\hat{s}+h\hat{t}\right) = ae+\hat{r}(af+be)+\hat{s}(ag+ce)+\hat{t}(ah+de)$$
$$+(bg+cf)\hat{r}\hat{s}+(bh+df)\hat{r}\hat{t}+(ch+dg)\hat{s}\hat{t}$$
$$+bf\hat{r}^2+cg\hat{s}^2+dh\hat{t}^2$$
$$=(ae+bh+cg+df)+(af+be+ch+dg)\hat{r}$$
$$+(ag+bf+ce+dh)\hat{s}+(ah+bg+cf+de)\hat{t}$$

Inspection of this leads to:

$$\left\{ \begin{array}{c} \hat{r}^2=\hat{s},\hat{s}^2=1,\hat{t}^2=\hat{s},\hat{r}\hat{s}=\hat{t},\hat{r}\hat{t}=1,\hat{s}\hat{t}=\hat{r} \\ \hat{s}=\sqrt[2]{+1},\hat{r}=\sqrt[4]{+1},\hat{t}=\sqrt[4]{+1} \end{array} \right\}$$

The form: $a+c\hat{s}$ is the 2-dimensional hyperbolic complex sub-algebra.

We have:

$$e^{\hat{r}\theta} = 1 + \hat{r}\theta + \frac{\hat{r}^2\theta^2}{2!} + \frac{\hat{r}^3\theta^3}{3!} + \frac{\hat{r}^4\theta^4}{4!} + \frac{\hat{r}^5\theta^5}{5!} + \frac{\hat{r}^6\theta^6}{6!} + \dots$$

$$= 1 + \hat{r}\theta + \frac{\hat{s}\theta^2}{2!} + \frac{\hat{t}\theta^3}{3!} + \frac{\theta^4}{4!} + \frac{\hat{r}\theta^5}{5!} + \frac{\hat{s}\theta^6}{6!} + \dots$$

$$= \left(1 + \frac{\theta^4}{4!} + \dots\right) + \hat{r}\left(\theta + \frac{\theta^5}{5!} + \dots\right) + \hat{s}\left(\frac{\theta^2}{2!} + \frac{\theta^6}{6!} + \dots\right) + \hat{t}\left(\frac{\theta^3}{3!} + \frac{\theta^7}{7!} + \dots\right)$$

$$= AH_4(\theta) + \hat{r}BH_4(\theta) + \hat{s}CH_4(\theta) + \hat{t}DH_4(\theta)$$

And:

$$e^{\hat{s}\theta} = 1 + \hat{s}\theta + \frac{\hat{s}^2\theta^2}{2!} + \frac{\hat{s}^3\theta^3}{3!} + \frac{\hat{s}^4\theta^4}{4!} + \frac{\hat{s}^5\theta^5}{5!} + \frac{\hat{s}^6\theta^6}{6!} + \dots$$

$$= 1 + \hat{s}\theta + \frac{\theta^2}{2!} + \frac{\hat{s}\theta^3}{3!} + \frac{\theta^4}{4!} + \frac{\hat{s}\theta^5}{5!} + \frac{\theta^6}{6!} + \dots$$

$$= \left(1 + \frac{\theta^2}{2!} + \frac{\theta^4}{4!} + \frac{\theta^6}{6!} + \dots\right) + \hat{s}\left(\theta + \frac{\theta^3}{3!} + \frac{\theta^5}{5!} + \dots\right)$$

$$= \cosh(\theta) + \hat{s}\sinh(\theta)$$

And:

$$e^{\hat{t}\theta} = 1 + \hat{t}\theta + \frac{\hat{t}^2\theta^2}{2!} + \frac{\hat{t}^3\theta^3}{3!} + \frac{\hat{t}^4\theta^4}{4!} + \frac{\hat{t}^5\theta^5}{5!} + \frac{\hat{t}^6\theta^6}{6!} + \dots$$

$$= 1 + \hat{t}\theta + \frac{\hat{s}\theta^2}{2!} + \frac{\hat{r}\theta^3}{3!} + \frac{\theta^4}{4!} + \frac{\hat{t}\theta^5}{5!} + \frac{\hat{s}\theta^6}{6!} + \dots$$

$$= \left(1 + \frac{\theta^4}{4!} + \dots\right) + \hat{t}\left(\theta + \frac{\theta^5}{5!} + \dots\right) + \hat{s}\left(\frac{\theta^2}{2!} + \frac{\theta^6}{6!} + \dots\right) + \hat{r}\left(\frac{\theta^3}{3!} + \dots\right)$$

$$= AH_4(\theta) + \hat{t}BH_4(\theta) + \hat{s}CH_4(\theta) + \hat{r}DH_4(\theta)$$

Leading to:

$$e^{b\hat{r} + c\hat{s} + d\hat{t}} = e^{b\hat{r}}e^{c\hat{s}}e^{d\hat{t}}$$

$$= PROD\begin{pmatrix} AH_4(b) + \hat{r}AH_4(b) + \hat{s}AH_4(b) + \hat{t}AH_4(b) \\ \cos(c) + \hat{s}\sin(c) \\ AH_4(d) + \hat{t}AH_4(d) + \hat{s}AH_4(d) + \hat{r}AH_4(d) \end{pmatrix}$$

$$= v_{[\,]}A(b,c,d) + \hat{r}v_{[\,]}B(b,c,d) + \hat{s}v_{[\,]}C(b,c,d) + \hat{t}v_{[\,]}D(b,c,d)$$

And:

$$e^{\frac{b\hat{r} + c\hat{s} + d\hat{t}}{n}} = v_{[\,]}A\left(\frac{b}{n},\frac{c}{n},\frac{d}{n}\right) + \hat{r}v_{[\,]}B\left(\frac{b}{n},\frac{c}{n},\frac{d}{n}\right) + \hat{s}v_{[\,]}C\left(\frac{b}{n},\frac{c}{n},\frac{d}{n}\right) + \hat{t}v_{[\,]}D\left(\frac{b}{n},\frac{c}{n},\frac{d}{n}\right)$$

Section 9 : Chapter 7

The Euclidean Shadow of the $C_{4.1}L^1H^3{}_{(j=1,k=1,l=1)}$ Algebra

In non-matrix notation, the $C_{4.1}L^1H^3{}_{(j=1,k=1,l=1)}$ algebra is the algebra of $a+b\hat{r}+c\hat{s}+d\hat{t}$. The multiplicative relations are:

$$\left\{\hat{r}\hat{s}=\hat{t},\hat{r}\hat{t}=1,\hat{s}\hat{t}=\hat{r},\hat{r}^2=\hat{s},\hat{s}^2=1,\hat{t}^2=\hat{s}\right\}:\hat{r}\neq\hat{s}\neq\hat{t}$$

The euclidean complex number 4th roots of unity have the same multiplicative relations with:

$$\left\{\hat{r}=\hat{i},\hat{s}=-1,\hat{t}=-\hat{i}\right\}$$

Substituting these euclidean complex roots into the non-matrix representation of the $C_{4.1}L^1H^3{}_{(j=1,k=1,l=1)}$ algebra, we get:

$$a+b\hat{r}+c\hat{s}+d\hat{t}=(a-c)+\hat{i}(b-d)$$

The polar form of this representation is:

$$\exp\left(\begin{bmatrix} a-c & b-d \\ -(b-d) & a-c \end{bmatrix}\right)=\begin{bmatrix} \dfrac{e^a}{e^c} & 0 \\ 0 & \dfrac{e^a}{e^c} \end{bmatrix}\begin{bmatrix} \cos(b-d) & \sin(b-d) \\ -\sin(b-d) & \cos(b-d) \end{bmatrix}$$

We see that the length within the 4-dimensional algebra, e^a, has now become $\dfrac{e^a}{e^c}$, and the 4-dimensional $\{b,c,d\}$ rotation matrices have been combined into a single 2-dimensional rotation matrix, but that the c rotation matrix is not represented in this rotation matrix.

The determinant of this euclidean complex number is:

$$(a-c)^2+(b-d)^2$$

Which the reader should compare to the determinant of the $C_{4.1}L^1H^3{}_{(j=1,k=1,l=1)}$ algebra which is:

$$\left((a-c)^2+(b-d)^2\right)\left((a+c)^2-(b+d)^2\right)$$

The factor $(a+c)^2-(b+d)^2$ is of the form of the 2-dimensional hyperbolic distance function – we seem to have folded out 2 hyperbolic dimensions.

Section 9 : Chapter 8

The Cauchy Riemann Equations - $C_{4.2}L^1H^3_{(j=1,k=1,l=1)}$

We have:

$$C_{4.2}L^1H^3_{(j=1,k=1,l=1)} = \begin{bmatrix} a & b & c & d \\ b & a & d & c \\ d & c & a & b \\ c & d & b & a \end{bmatrix}$$

Differentiating this algebra gives:

$$\frac{\partial \begin{bmatrix} f(a,b,c,d) & g(a,b,c,d) & h(a,b,c,d) & i(a,b,c,d) \\ g(a,b,c,d) & f(a,b,c,d) & i(a,b,c,d) & h(a,b,c,d) \\ i(a,b,c,d) & h(a,b,c,d) & f(a,b,c,d) & g(a,b,c,d) \\ h(a,b,c,d) & i(a,b,c,d) & g(a,b,c,d) & f(a,b,c,d) \end{bmatrix}}{\partial \begin{bmatrix} a & 0 & 0 & 0 \\ 0 & a & 0 & 0 \\ 0 & 0 & a & 0 \\ 0 & 0 & 0 & a \end{bmatrix}} = \begin{bmatrix} \frac{\partial f}{\partial a} & \frac{\partial g}{\partial a} & \frac{\partial h}{\partial a} & \frac{\partial i}{\partial a} \\ \frac{\partial g}{\partial a} & \frac{\partial f}{\partial a} & \frac{\partial i}{\partial a} & \frac{\partial h}{\partial a} \\ \frac{\partial i}{\partial a} & \frac{\partial h}{\partial a} & \frac{\partial f}{\partial a} & \frac{\partial g}{\partial a} \\ \frac{\partial h}{\partial a} & \frac{\partial i}{\partial a} & \frac{\partial g}{\partial a} & \frac{\partial f}{\partial a} \end{bmatrix}$$

$$\frac{\partial \begin{bmatrix} f(a,b,c,d) & g(a,b,c,d) & h(a,b,c,d) & i(a,b,c,d) \\ g(a,b,c,d) & f(a,b,c,d) & i(a,b,c,d) & h(a,b,c,d) \\ i(a,b,c,d) & h(a,b,c,d) & f(a,b,c,d) & g(a,b,c,d) \\ h(a,b,c,d) & i(a,b,c,d) & g(a,b,c,d) & f(a,b,c,d) \end{bmatrix}}{\partial \begin{bmatrix} 0 & b & 0 & 0 \\ b & 0 & 0 & 0 \\ 0 & 0 & 0 & b \\ 0 & 0 & b & 0 \end{bmatrix}} = \begin{bmatrix} \frac{\partial g}{\partial b} & \frac{\partial f}{\partial b} & \frac{\partial i}{\partial b} & \frac{\partial h}{\partial b} \\ \frac{\partial f}{\partial b} & \frac{\partial g}{\partial b} & \frac{\partial h}{\partial b} & \frac{\partial i}{\partial b} \\ \frac{\partial h}{\partial b} & \frac{\partial i}{\partial b} & \frac{\partial g}{\partial b} & \frac{\partial f}{\partial b} \\ \frac{\partial i}{\partial b} & \frac{\partial h}{\partial b} & \frac{\partial f}{\partial b} & \frac{\partial g}{\partial b} \end{bmatrix}$$

$$\frac{\partial \begin{bmatrix} f(a,b,c,d) & g(a,b,c,d) & h(a,b,c,d) & i(a,b,c,d) \\ g(a,b,c,d) & f(a,b,c,d) & i(a,b,c,d) & h(a,b,c,d) \\ i(a,b,c,d) & h(a,b,c,d) & f(a,b,c,d) & g(a,b,c,d) \\ h(a,b,c,d) & i(a,b,c,d) & g(a,b,c,d) & f(a,b,c,d) \end{bmatrix}}{\partial \begin{bmatrix} 0 & 0 & c & 0 \\ 0 & 0 & 0 & c \\ 0 & c & 0 & 0 \\ c & 0 & 0 & 0 \end{bmatrix}} = \begin{bmatrix} \frac{\partial h}{\partial c} & \frac{\partial i}{\partial c} & \frac{\partial g}{\partial c} & \frac{\partial f}{\partial c} \\ \frac{\partial i}{\partial c} & \frac{\partial h}{\partial c} & \frac{\partial f}{\partial c} & \frac{\partial g}{\partial c} \\ \frac{\partial f}{\partial c} & \frac{\partial g}{\partial c} & \frac{\partial h}{\partial c} & \frac{\partial i}{\partial c} \\ \frac{\partial g}{\partial c} & \frac{\partial f}{\partial c} & \frac{\partial i}{\partial c} & \frac{\partial h}{\partial c} \end{bmatrix}$$

$$\frac{\partial \begin{bmatrix} f(a,b,c,d) & g(a,b,c,d) & h(a,b,c,d) & i(a,b,c,d) \\ g(a,b,c,d) & f(a,b,c,d) & i(a,b,c,d) & h(a,b,c,d) \\ i(a,b,c,d) & h(a,b,c,d) & f(a,b,c,d) & g(a,b,c,d) \\ h(a,b,c,d) & i(a,b,c,d) & g(a,b,c,d) & f(a,b,c,d) \end{bmatrix}}{\partial \begin{bmatrix} 0 & 0 & 0 & d \\ 0 & 0 & d & 0 \\ d & 0 & 0 & 0 \\ 0 & d & 0 & 0 \end{bmatrix}} = \begin{bmatrix} \dfrac{\partial i}{\partial d} & \dfrac{\partial h}{\partial d} & \dfrac{\partial f}{\partial d} & \dfrac{\partial g}{\partial d} \\[2mm] \dfrac{\partial h}{\partial d} & \dfrac{\partial i}{\partial d} & \dfrac{\partial g}{\partial d} & \dfrac{\partial f}{\partial d} \\[2mm] \dfrac{\partial g}{\partial d} & \dfrac{\partial f}{\partial d} & \dfrac{\partial i}{\partial d} & \dfrac{\partial h}{\partial d} \\[2mm] \dfrac{\partial f}{\partial d} & \dfrac{\partial g}{\partial d} & \dfrac{\partial h}{\partial d} & \dfrac{\partial i}{\partial d} \end{bmatrix}$$

From this, the Cauchy Riemann equations of the $C_{4.2}L^1H^3{}_{(j=1,k=1,l=1)}$ algebra are:

$$\frac{\partial f}{\partial a} = \frac{\partial g}{\partial b} = \frac{\partial h}{\partial c} = \frac{\partial i}{\partial d}$$

$$\frac{\partial g}{\partial a} = \frac{\partial f}{\partial b} = \frac{\partial i}{\partial c} = \frac{\partial h}{\partial d}$$

$$\frac{\partial h}{\partial a} = \frac{\partial i}{\partial b} = \frac{\partial g}{\partial c} = \frac{\partial f}{\partial d}$$

$$\frac{\partial i}{\partial a} = \frac{\partial h}{\partial b} = \frac{\partial f}{\partial c} = \frac{\partial g}{\partial d}$$

The differential is:

$$d\begin{bmatrix} f(a,b,c,d) & g(a,b,c,d) & h(a,b,c,d) & i(a,b,c,d) \\ g(a,b,c,d) & f(a,b,c,d) & i(a,b,c,d) & h(a,b,c,d) \\ i(a,b,c,d) & h(a,b,c,d) & f(a,b,c,d) & g(a,b,c,d) \\ h(a,b,c,d) & i(a,b,c,d) & g(a,b,c,d) & f(a,b,c,d) \end{bmatrix} =$$

$$\begin{bmatrix} \dfrac{\partial f}{\partial a}+\dfrac{\partial g}{\partial b}+\dfrac{\partial h}{\partial c}+\dfrac{\partial i}{\partial d} & \dfrac{\partial g}{\partial a}+\dfrac{\partial f}{\partial b}+\dfrac{\partial i}{\partial c}+\dfrac{\partial h}{\partial d} & \dfrac{\partial h}{\partial a}+\dfrac{\partial i}{\partial b}+\dfrac{\partial g}{\partial c}+\dfrac{\partial f}{\partial d} & \dfrac{\partial i}{\partial a}+\dfrac{\partial h}{\partial b}+\dfrac{\partial f}{\partial c}+\dfrac{\partial g}{\partial d} \\[2mm] \sim & \sim & \sim & \sim \\[1mm] \sim & \sim & \sim & \sim \\[1mm] \sim & \sim & \sim & \sim \end{bmatrix}$$

where we have twiddled out the duplicate entries.

Introduction to Section 10

In this section, we look more closely at the 4-dimensional simple trigonometric functions and some of the 4-dimensional ν-functions. The reader is pointed toward the graphs of the simple trigonometric functions. This section contains no great insights, but it will familiarise the reader with 4-dimensional trigonometry.

Section 10 : Chapter 1

The 4-dimensional Simple Trigonometric Functions

The 4-dimensional simple trigonometric functions are defined as 4-way splittings of the exponential series.

$$AH_4(x) = hypergeom\left([\ ],\left[\frac{1}{4},\frac{1}{2},\frac{3}{4}\right],\left(\frac{x}{4}\right)^4\right) = 1 + \frac{x^4}{4!} + \frac{x^8}{8!} + \frac{x^{12}}{12!} + ...$$

$$BH_4(x) = x\left(hypergeom\left([\ ],\left[\frac{1}{2},\frac{3}{4},\frac{5}{4}\right],\left(\frac{x}{4}\right)^4\right)\right) = x + \frac{x^5}{5!} + \frac{x^9}{9!} + \frac{x^{13}}{13!} + ...$$

$$CH_4(x) = \frac{x^2}{2}\left(hypergeom\left([\ ],\left[\frac{3}{4},\frac{5}{4},\frac{6}{4}\right],\left(\frac{x}{4}\right)^4\right)\right) = \frac{x^2}{2!} + \frac{x^6}{6!} + \frac{x^{10}}{10!} + ...$$

$$DH_4(x) = \frac{x^3}{6}\left(hypergeom\left([\ ],\left[\frac{5}{4},\frac{6}{4},\frac{7}{4}\right],\left(\frac{x}{4}\right)^4\right)\right) = \frac{x^3}{3!} + \frac{x^7}{7!} + \frac{x^{11}}{11!} + ...$$

$$AE_4(x) = hypergeom\left([\ ],\left[\frac{1}{4},\frac{1}{2},\frac{3}{4}\right],-\left(\frac{x}{4}\right)^4\right) = 1 - \frac{x^4}{4!} + \frac{x^8}{8!} - \frac{x^{12}}{12!} + ...$$

$$BE_4(x) = x\left(hypergeom\left([\ ],\left[\frac{1}{2},\frac{3}{4},\frac{5}{4}\right],-\left(\frac{x}{4}\right)^4\right)\right) = x - \frac{x^5}{5!} + \frac{x^9}{9!} - \frac{x^{13}}{13!} + ...$$

$$CE_4(x) = \frac{x^2}{2}\left(hypergeom\left([\ ],\left[\frac{3}{4},\frac{5}{4},\frac{6}{4}\right],-\left(\frac{x}{4}\right)^4\right)\right) = \frac{x^2}{2!} - \frac{x^6}{6!} + \frac{x^{10}}{10!} - ...$$

$$DE_4(x) = \frac{x^3}{6}\left(hypergeom\left([\ ],\left[\frac{5}{4},\frac{6}{4},\frac{7}{4}\right],-\left(\frac{x}{4}\right)^4\right)\right) = \frac{x^3}{3!} - \frac{x^7}{7!} + \frac{x^{11}}{11!} - ...$$

Clearly,

$$AH_4(x) + BH_4(x) + CH_4(x) + DH_4(x) = e^x$$

$$AH_4(x) + CH_4(x) = \cosh(x)$$

$$BH_4(x) + DH_4(x) = \sinh(x)$$

$$AH_4(x) - CH_4(x) = \cos(x)$$

$$BH_4(x) - DH_4(x) = \sin(x)$$

142

As with all simple trigonometric functions, these have a differentiation cycle.[70] In standard function form, the 4-dimensional H-type simple trigonometric functions are:

$$AH_4(x) = \frac{1}{4}(2\cosh x + 2\cos x)$$

$$BH_4(x) = \frac{1}{4}(2\sinh x + 2\sin x)$$

$$CH_4(x) = \frac{1}{4}(2\cosh x - 2\cos x)$$

$$DH_4(x) = \frac{1}{4}(2\sinh x - 2\sin x)$$

The graphs of the 4-dimensional H-type simple trigonometric functions are:

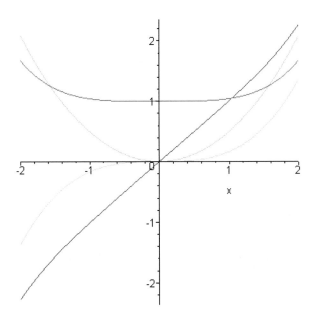

[70] As is clear from examination of the series expansions.

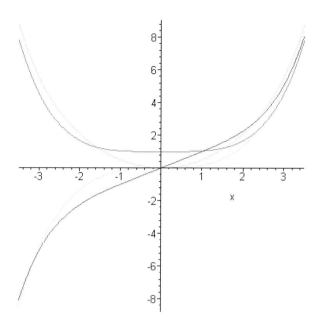

These functions are not periodic.

The 4-dimensional E-type simple trigonometric functions are:

$$AE_4(x) = \cos\left(\frac{x}{\sqrt{2}}\right)\cosh\left(\frac{x}{\sqrt{2}}\right)$$

$$BE_4(x) = \frac{1}{\sqrt{2}}\left(\cos\left(\frac{x}{\sqrt{2}}\right)\sinh\left(\frac{x}{\sqrt{2}}\right) + \cosh\left(\frac{x}{\sqrt{2}}\right)\sin\left(\frac{x}{\sqrt{2}}\right)\right)$$

$$CE_4(x) = \sin\left(\frac{x}{\sqrt{2}}\right)\sinh\left(\frac{x}{\sqrt{2}}\right)$$

$$DE_4(x) = \frac{1}{\sqrt{2}}\left(\sin\left(\frac{x}{\sqrt{2}}\right)\cosh\left(\frac{x}{\sqrt{2}}\right) - \cos\left(\frac{x}{\sqrt{2}}\right)\sinh\left(\frac{x}{\sqrt{2}}\right)\right)$$

144

The graphs of the 4-dimensional E-type simple trigonometric functions are:

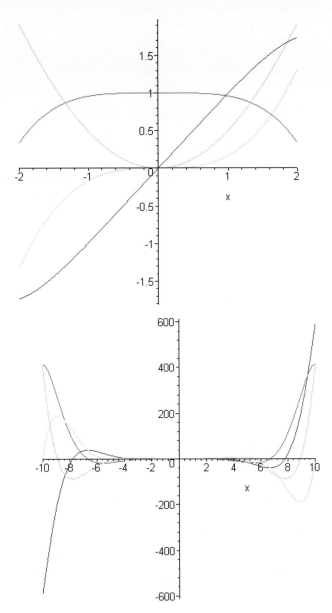

These functions are wave-like with enormous amplitude for large positive and large negative values of x; they are periodic.

Section 10 : Chapter 2

The v-functions - $C_{4.1}L^1H^3{}_{(j=1,k=1,l=1)}$

The simple polar form of the $C_{4.1}L^1H^3{}_{(j=1,k=1,l=1)}$ algebra is:

$$\exp\left(\begin{bmatrix} a & b & c & d \\ d & a & b & c \\ c & d & a & b \\ b & c & d & a \end{bmatrix}\right) = PROD \left(\begin{bmatrix} e^a & 0 & 0 & 0 \\ 0 & e^a & 0 & 0 \\ 0 & 0 & e^a & 0 \\ 0 & 0 & 0 & e^a \end{bmatrix} \begin{bmatrix} AH_4(b) & BH_4(b) & CH_4(b) & DH_4(b) \\ DH_4(b) & AH_4(b) & BH_4(b) & CH_4(b) \\ CH_4(b) & DH_4(b) & AH_4(b) & BH_4(b) \\ BH_4(b) & CH_4(b) & DH_4(b) & AH_4(b) \end{bmatrix} \begin{bmatrix} \cosh(c) & 0 & \sinh(c) & 0 \\ 0 & \cosh(c) & 0 & \sinh(c) \\ \sinh(c) & 0 & \cosh(c) & 0 \\ 0 & \sinh(c) & 0 & \cosh(c) \end{bmatrix} \begin{bmatrix} AH_4(d) & DH_4(d) & CH_4(d) & BH_4(d) \\ BH_4(d) & AH_4(d) & DH_4(d) & CH_4(d) \\ CH_4(d) & BH_4(d) & AH_4(d) & DH_4(d) \\ DH_4(d) & CH_4(d) & BH_4(d) & AH_4(d) \end{bmatrix} \right)$$

Multiplying the three rotation matrices together gives the compound rotation matrix containing the 4-dimensional v-functions of this algebra:

$$\begin{bmatrix} v_{[\,]}A(b,c,d) & v_{[\,]}B(b,c,d) & v_{[\,]}C(b,c,d) & v_{[\,]}D(b,c,d) \\ v_{[\,]}D(b,c,d) & v_{[\,]}A(b,c,d) & v_{[\,]}B(b,c,d) & v_{[\,]}C(b,c,d) \\ v_{[\,]}C(b,c,d) & v_{[\,]}D(b,c,d) & v_{[\,]}A(b,c,d) & v_{[\,]}B(b,c,d) \\ v_{[\,]}B(b,c,d) & v_{[\,]}C(b,c,d) & v_{[\,]}D(b,c,d) & v_{[\,]}A(b,c,d) \end{bmatrix}$$

We have:

$$v_{\left[C_{4.1}L^1H^3{}_{(j=1,k=1,l=1)}\right]}A(b,c,d) = \frac{1}{4}\left(2e^c\cosh(b+d)+2e^{-c}\cos(b-d)\right)$$

$$v_{\left[C_{4.1}L^1H^3{}_{(j=1,k=1,l=1)}\right]}B(b,c,d) = \frac{1}{4}\left(2e^c\sinh(b+d)+2e^{-c}\sin(b-d)\right)$$

$$V_{\left[C_{4,1}L^1H^3_{(j=1,k=1,l=1)}\right]}C(b,c,d)=\frac{1}{4}\left(2e^c\cosh(b+d)-2e^{-c}\cos(b-d)\right)$$

$$V_{\left[C_{4,1}L^1H^3_{(j=1,k=1,l=1)}\right]}D(b,c,d)=\frac{1}{4}\left(2e^c\sinh(b+d)-2e^{-c}\sin(b-d)\right)$$

We have:

$$v_{[\ ]}A(b,c,d)+v_{[\ ]}B(b,c,d)+v_{[\ ]}C(b,c,d)+v_{[\ ]}D(b,c,d)=e^{b+c+d}$$

The differentiation cycles are:

$$\frac{\partial\left(v_{[\ ]}A(b,c,d)\right)}{\partial b}=v_{[\ ]}D(b,c,d)\qquad\qquad\frac{\partial\left(v_{[\ ]}B(b,c,d)\right)}{\partial b}=v_{[\ ]}A(b,c,d)$$

$$\frac{\partial\left(v_{[\ ]}A(b,c,d)\right)}{\partial c}=v_{[\ ]}C(b,c,d)\qquad\qquad\frac{\partial\left(v_{[\ ]}B(b,c,d)\right)}{\partial c}=v_{[\ ]}D(b,c,d)$$

$$\frac{\partial\left(v_{[\ ]}A(b,c,d)\right)}{\partial d}=v_{[\ ]}B(b,c,d)\qquad\qquad\frac{\partial\left(v_{[\ ]}B(b,c,d)\right)}{\partial d}=v_{[\ ]}C(b,c,d)$$

$$\frac{\partial\left(v_{[\ ]}C(b,c,d)\right)}{\partial b}=v_{[\ ]}B(b,c,d)\qquad\qquad\frac{\partial\left(v_{[\ ]}D(b,c,d)\right)}{\partial b}=v_{[\ ]}C(b,c,d)$$

$$\frac{\partial\left(v_{[\ ]}C(b,c,d)\right)}{\partial c}=v_{[\ ]}A(b,c,d)\qquad\qquad\frac{\partial\left(v_{[\ ]}D(b,c,d)\right)}{\partial c}=v_{[\ ]}B(b,c,d)$$

$$\frac{\partial\left(v_{[\ ]}C(b,c,d)\right)}{\partial d}=v_{[\ ]}D(b,c,d)\qquad\qquad\frac{\partial\left(v_{[\ ]}D(b,c,d)\right)}{\partial d}=v_{[\ ]}A(b,c,d)$$

The $C_{4,1}L^1H^3_{(j=1,k=1,l=1)}$ v-functions have series expansions. We list the case of $v_{[\ ]}A$:
Expanded about $b=0$:

$$v_{[\ ]}A(b,c,d)=v_{[\ ]}A(0,c,d)+v_{[\ ]}B(0,c,d)b+v_{[\ ]}C(0,c,d)\frac{b^2}{2!}$$

$$+v_{[\ ]}D(0,c,d)\frac{b^3}{3!}+v_{[\ ]}A(0,c,d)\frac{b^4}{4!}+...$$

This series is a sum of all four v-functions.

Expanded about $c=0$:

$$v_{[\ ]}A(b,c,d)=v_{[\ ]}A(b,0,d)+v_{[\ ]}C(b,0,d)c+v_{[\ ]}A(b,0,d)\frac{c^2}{2!}$$

$$+v_{[\ ]}C(b,0,d)\frac{c^3}{3!}+v_{[\ ]}A(b,0,d)\frac{c^4}{4!}+...$$

This series is a sum of only two v-functions.

Expanded about $d = 0$:

$$V_{[\ ]}A(b,c,d) = V_{[\ ]}A(b,c,0) + V_{[\ ]}B(b,c,0)d + V_{[\ ]}C(b,c,0)\frac{d^2}{2!}$$

$$+ V_{[\ ]}D(b,c,0)\frac{d^3}{3!} + V_{[\ ]}A(b,c,0)\frac{d^4}{4!} + ...$$

This series is a sum of all four v-functions.

The series form of the 3-dimensional v-functions, is a sum of the 3-dimensional simple trigonometric functions. However, in 3-dimensions, the v-functions with one argument equal to zero are the simple trigonometric functions. Thus, we could equally well have written the 3-dimensional series with v-functions with one argument zero instead of the simple trigonometric functions. Since the v-functions have a differentiation cycle and the series are calculated as a sum of the differentials, this is not surprising.

Section 10 : Addendum 1 to Chapter 2

The v-functions of the $C_{4.2}L^1H^3_{(j=1,k=1,l=1)}$ alebra are:

$$V_{\left[C_{4.2}L^1H^3_{(j=1,k=1,l=1)}\right]}A(b,c,d) = \frac{1}{4}\left(2e^b\cosh(c+d) + 2e^{-b}\cos(c-d)\right)$$

$$V_{\left[C_{4.2}L^1H^3_{(j=1,k=1,l=1)}\right]}B(b,c,d) = \frac{1}{4}\left(2e^b\cosh(c+d) - 2e^{-b}\cos(c-d)\right)$$

$$V_{\left[C_{4.2}L^1H^3_{(j=1,k=1,l=1)}\right]}C(b,c,d) = \frac{1}{4}\left(2e^b\sinh(c+d) - 2e^{-b}\sin(c-d)\right)$$

$$V_{\left[C_{4.2}L^1H^3_{(j=1,k=1,l=1)}\right]}D(b,c,d) = \frac{1}{4}\left(2e^b\sinh(c+d) + 2e^{-b}\sin(c-d)\right)$$

which are effectively the same (variables have swapped about and names have swapped about) as the v-functions of the $C_{4.1}L^1H^3_{(j=1,k=1,l=1)}$ algebra.

The v-functions of the $C_{4.1}L^1E^3_{(j=-1,k=1,l=1)}$ algebra are:

$$V_{\left[C_{4.1}L^1E^3_{(j=-1,k=1,l=1)}\right]}A(b,c,d) = \frac{1}{2}\left(e^{\left(\frac{b-d}{\sqrt{2}}\right)}\cos\left(\frac{b+d}{\sqrt{2}}+c\right) + e^{-\left(\frac{b-d}{\sqrt{2}}\right)}\cos\left(\frac{b+d}{\sqrt{2}}-c\right)\right)$$

$$V_{\left[C_{4.1}L^1E^3_{(j=-1,k=1,l=1)}\right]}B(b,c,d) = \frac{\sqrt{2}}{4}\left(\begin{array}{l} e^{\left(\frac{b-d}{\sqrt{2}}\right)}\left(\cos\left(\frac{b+d}{\sqrt{2}}+c\right) + \sin\left(\frac{b+d}{\sqrt{2}}+c\right)\right) \\ + e^{-\left(\frac{b-d}{\sqrt{2}}\right)}\left(-\cos\left(\frac{b+d}{\sqrt{2}}-c\right) + \sin\left(\frac{b+d}{\sqrt{2}}-c\right)\right) \end{array}\right)$$

$$V_{\left[C_{4,1}L^1E^3_{(j=-1,k=1,l=1)}\right]}C(b,c,d)=\frac{1}{2}\left(e^{\left(\frac{b-d}{\sqrt{2}}\right)}\sin\left(\frac{b+d}{\sqrt{2}}+c\right)-e^{-\left(\frac{b-d}{\sqrt{2}}\right)}\sin\left(\frac{b+d}{\sqrt{2}}-c\right)\right)$$

$$V_{\left[C_{4,1}L^1E^3_{(j=-1,k=1,l=1)}\right]}D(b,c,d)=\frac{\sqrt{2}}{4}\left(\begin{array}{l}e^{\left(\frac{b-d}{\sqrt{2}}\right)}\left(-\cos\left(\frac{b+d}{\sqrt{2}}+c\right)+\sin\left(\frac{b+d}{\sqrt{2}}+c\right)\right)\\[2ex]+e^{-\left(\frac{b-d}{\sqrt{2}}\right)}\left(\cos\left(\frac{b+d}{\sqrt{2}}-c\right)+\sin\left(\frac{b+d}{\sqrt{2}}-c\right)\right)\end{array}\right)$$

As with the v-functions of the other algebras we have met, the v-functions of these algebras also have series expansions which have v-functions for the coefficients.

Introduction to Section 11

In this section, we explore the connection between the natural algebras and the abelian groups. We are already aware that the algebraic matrix forms are copies of the standard form Cayley tables of the abelian groups. In this section, we find out that, in a profound sense, they are the abelian groups. What we have here is a unification of abelian groups and the natural algebraic fields. Since the natural algebraic fields contain the natural spaces, this is a unification of abelian groups, natural algebraic fields, and natural geometries.

The unification is two-fold more extensive than this. Firstly, the natural algebraic fields are the vector product algebras. Secondly, we have abandoned the concept of algebraic extension of fields in favour of algebraic inclusion. This is effectively declaring that each of the natural algebraic fields is a different type of number. Thus we have it that each type of number is also an abelian group and also a natural algebra and also a vector product algebra and also a natural space.

Section 11 : Chapter 1

Permutation Matrices

Square matrices whose elements are either 1 or 0 and in which these elements are arranged so that a 1 occurs once and only once in each row and in each column are called permutation matrices. We give some examples of 4×4 permutation matrices.

$$A:\begin{pmatrix}1234\\2341\end{pmatrix}=(1234)\mapsto\begin{bmatrix}0&1&0&0\\0&0&1&0\\0&0&0&1\\1&0&0&0\end{bmatrix}$$

$$B:\begin{pmatrix}1234\\3421\end{pmatrix}=(1324)\mapsto\begin{bmatrix}0&0&1&0\\0&0&0&1\\0&1&0&0\\1&0&0&0\end{bmatrix}$$

$$C:\begin{pmatrix}1234\\2413\end{pmatrix}=(1243)\mapsto\begin{bmatrix}0&1&0&0\\0&0&0&1\\1&0&0&0\\0&0&1&0\end{bmatrix}$$

$$D:\begin{pmatrix}1234\\4132\end{pmatrix}=(142)(3)\mapsto\begin{bmatrix}0&0&0&1\\1&0&0&0\\0&0&1&0\\0&1&0&0\end{bmatrix}$$

$$E:\begin{pmatrix}1234\\2143\end{pmatrix}=(12)(34)\mapsto\begin{bmatrix}0&1&0&0\\1&0&0&0\\0&0&0&1\\0&0&1&0\end{bmatrix}$$

Convention has it that all of the above matrices are permutation matrices of equal status. We differ in this view and we refer to the unbroken permutations $\{A,B,C\}$ as pure permutations and the broken permutations as impure permutations.

Nomenclature:

Pure Permutation Matrices

A pure permutation matrix is a permutation matrix that is associated with an unbroken permutation cycle.

Permutation matrices are made by swapping columns of the identity matrix. Since the identity matrix is non-singular and non-singularity is maintained under column swaps, permutation matrices are always non-singular. The inverse of a permutation matrix corresponds to the inverse permutation.

Since every finite group can be represented as a permutation group, it is of no great surprise that permutation matrices are associated with groups. The unbroken permutations above, $\{A,B,C\}$, are associated with the cyclic group, C_4.

The broken permutation, D, is associated with 'fixing together' the groups $\{C_1,C_3\}$.

$$\begin{bmatrix} 0 & 0 & 0 & 1 \\ 1 & 0 & 0 & 0 \\ 0 & 0 & 1 & 0 \\ 0 & 1 & 0 & 0 \end{bmatrix} \mapsto \begin{bmatrix} \begin{bmatrix} 0 & 1 & 0 \\ 0 & 0 & 1 \\ 1 & 0 & 0 \end{bmatrix} & 0 \\ 0 & 0 & 0 & \begin{bmatrix} 1 \end{bmatrix} \end{bmatrix}$$

The presence of a 1 on the leading diagonal betrays the presence of the C_1 group. Of course, 'fixing together' of groups is an impure thing. What we have here is the 3×3 matrix of a pure permutation and the 1×1 matrix of a pure permutation fitted on to the leading diagonal of a 4×4 matrix. If we were to form the direct product of the groups $\{C_1,C_3\}$, we would get the group C_3, and this is associated with 3×3 permutation matrices. Such 'fixed together' permutation matrices are not our concern.

All the permutation matrices generate (by matrix multiplication – we raise them to various powers of themselves) the group with which they are associated. We demonstrate with the 1×1 permutation matrices, the 2×2 permutation matrices, and the 3×3 permutation matrices. The powers of these matrices are:

$$[1]^1 = [1]$$

$$\begin{bmatrix} 1 & 0 \\ 0 & 1 \end{bmatrix}^1 = \begin{bmatrix} 1 & 0 \\ 0 & 1 \end{bmatrix}$$

$$\begin{bmatrix} 0 & 1 \\ 1 & 0 \end{bmatrix}^{1..2} = \begin{bmatrix} 0 & 1 \\ 1 & 0 \end{bmatrix}, \begin{bmatrix} 1 & 0 \\ 0 & 1 \end{bmatrix}$$

$$F: \begin{bmatrix} 1 & 0 & 0 \\ 0 & 1 & 0 \\ 0 & 0 & 1 \end{bmatrix}^1 = \begin{bmatrix} 1 & 0 & 0 \\ 0 & 1 & 0 \\ 0 & 0 & 1 \end{bmatrix} : C_1$$

$$G: \begin{bmatrix} 1 & 0 & 0 \\ 0 & 0 & 1 \\ 0 & 1 & 0 \end{bmatrix}^{1..2} = \begin{bmatrix} 1 & 0 & 0 \\ 0 & 0 & 1 \\ 0 & 1 & 0 \end{bmatrix}, \begin{bmatrix} 1 & 0 & 0 \\ 0 & 1 & 0 \\ 0 & 0 & 1 \end{bmatrix} : C_2 \times C_1$$

$$H: \begin{bmatrix} 0 & 1 & 0 \\ 1 & 0 & 0 \\ 0 & 0 & 1 \end{bmatrix}^{1..2} = \begin{bmatrix} 0 & 1 & 0 \\ 1 & 0 & 0 \\ 0 & 0 & 1 \end{bmatrix}, \begin{bmatrix} 1 & 0 & 0 \\ 0 & 1 & 0 \\ 0 & 0 & 1 \end{bmatrix} : C_2 \times C_1$$

$$I: \begin{bmatrix} 0 & 1 & 0 \\ 0 & 0 & 1 \\ 1 & 0 & 0 \end{bmatrix}^{1..3} = \begin{bmatrix} 0 & 1 & 0 \\ 0 & 0 & 1 \\ 1 & 0 & 0 \end{bmatrix}, \begin{bmatrix} 0 & 0 & 1 \\ 1 & 0 & 0 \\ 0 & 1 & 0 \end{bmatrix}, \begin{bmatrix} 1 & 0 & 0 \\ 0 & 1 & 0 \\ 0 & 0 & 1 \end{bmatrix} : C_3$$

$$J: \begin{bmatrix} 0 & 0 & 1 \\ 1 & 0 & 0 \\ 0 & 1 & 0 \end{bmatrix}^{1..3} = \begin{bmatrix} 0 & 0 & 1 \\ 1 & 0 & 0 \\ 0 & 1 & 0 \end{bmatrix}, \begin{bmatrix} 0 & 1 & 0 \\ 0 & 0 & 1 \\ 1 & 0 & 0 \end{bmatrix}, \begin{bmatrix} 1 & 0 & 0 \\ 0 & 1 & 0 \\ 0 & 0 & 1 \end{bmatrix} : C_3$$

$$K: \begin{bmatrix} 0 & 0 & 1 \\ 0 & 1 & 0 \\ 1 & 0 & 0 \end{bmatrix}^{1..2} = \begin{bmatrix} 0 & 0 & 1 \\ 0 & 1 & 0 \\ 1 & 0 & 0 \end{bmatrix}, \begin{bmatrix} 1 & 0 & 0 \\ 0 & 1 & 0 \\ 0 & 0 & 1 \end{bmatrix} : C_2 \times C_1$$

Of all the 3×3 permutation matrices, only $\{I, J\}$ are pure permutation matrices. The other matrices are 3×3 matrices with lesser matrices fitted inside them along the leading diagonal.

In the case of 3×3 permutation matrices, we select the three powers of either of the pure permutation matrices $\{I, J\}$ - they generate the same matrices.

$$I^1 = \begin{bmatrix} 0 & 1 & 0 \\ 0 & 0 & 1 \\ 1 & 0 & 0 \end{bmatrix}, I^2 = \begin{bmatrix} 0 & 0 & 1 \\ 1 & 0 & 0 \\ 0 & 1 & 0 \end{bmatrix}, I^3 = \begin{bmatrix} 1 & 0 & 0 \\ 0 & 1 & 0 \\ 0 & 0 & 1 \end{bmatrix}$$

We change them into variable matrices.

$$B = \begin{bmatrix} 0 & b & 0 \\ 0 & 0 & b \\ b & 0 & 0 \end{bmatrix}, C = \begin{bmatrix} 0 & 0 & c \\ c & 0 & 0 \\ 0 & c & 0 \end{bmatrix}, A = \begin{bmatrix} a & 0 & 0 \\ 0 & a & 0 \\ 0 & 0 & a \end{bmatrix}$$

We add them.

$$\begin{bmatrix} a & b & c \\ c & a & b \\ b & c & a \end{bmatrix} = C_3 L^j H^2_{\;(j=1,k=1)}$$

Thus, from the set of 3×3 pure permutation matrices, we have constructed the basic layout of the 3-dimensional algebraic matrix form[71].

In the case of the 4×4 permutation matrices, there are nine permutation matrices that have only zeros on the leading diagonal. They come in sets of threes (two C_4s and one C_2[72]) which, when turned into variables and added, produce the three 4-dimensional algebraic matrix forms $\{C_{4.1}L^j H^3, C_{4.2}L^j H^3, C_{4.3}L^j H^3\}$, and so these algebras are associated with the group C_4.

There is one other way of combining three of these nine (all zeros on the leading diagonal) 4×4 permutation matrices; the three C_2 permutation matrices can be combined together, and this gives the $C_2 \times C_2 L^j H^3$ algebraic matrix form, and so this algebra is thus associated with the only other abelian group of order 4, $C_2 \times C_2$.

The group C_4 has three different, but algebraically isomorphic, 4×4 algebraic matrix forms associated with it. In one of these, $C_{4.1}L^j H^3$, the C_2 sub-group is generated by the c element; in the $C_{4.2}L^j H^3$ algebra, the C_2 sub-group is generated by the b element; in the $C_{4.3}L^j H^3$ algebra, the C_2 sub-group is generated by the d element.

Perusal of the nine 4×4 permutation matrices that have only zeros on the leading diagonal will give the reader a clearer understanding of the basis of algebraic isomorphisms within these algebras.

We have now established a clear connection between the three and four dimensional algebraic matrix forms and the finite abelian groups of order three and four. The connection expresses itself very succinctly in the Cayley tables of the finite abelian groups. For example, the standard form Cayley table of the group C_3 is:

A	B	C
C	A	B
B	C	A

which, if copied into a matrix, forms the $C_3 L^j H^2$ algebraic matrix form:

$$C_3 L^j H^2 = \begin{bmatrix} a & b & c \\ c & a & b \\ b & c & a \end{bmatrix}$$

---

[71] And the Cayley table of the order 3 cyclic group, of course.

[72] C_4 has a C_2 sub-group.

This phenomenon is general, and we can write down the algebraic matrix form associated with any finite abelian group by copying the standard form Cayley table of that group into a matrix. In short, we are associating an algebra with every finite abelian group.

Roots of Unity:

The eigenvalues of a permutation matrix are the roots of unity associated with the size of the sub-matrices. For example consider the permutation matrix:

$$P_\sigma = \begin{bmatrix} 0 & 0 & 0 & 0 & 1 & 0 \\ 1 & 0 & 0 & 0 & 0 & 0 \\ 0 & 0 & 0 & 0 & 0 & 1 \\ 0 & 0 & 0 & 1 & 0 & 0 \\ 0 & 1 & 0 & 0 & 0 & 0 \\ 0 & 0 & 1 & 0 & 0 & 0 \end{bmatrix} \equiv \begin{bmatrix} \begin{bmatrix} 0 & 1 & 0 \\ 0 & 0 & 1 \\ 1 & 0 & 0 \end{bmatrix} & 0 & 0 \\ 0 & [1] & 0 \\ 0 & 0 & \begin{bmatrix} 0 & 1 \\ 1 & 0 \end{bmatrix} \end{bmatrix}$$

The characteristic polynomial of this matrix is: $(\lambda^3 - 1)(\lambda - 1)(\lambda^2 - 1) = 0$, and so it has

eigenvalues $\left\{ 1^{\frac{1}{3}}, 1^{\frac{1}{3}}, 1^{\frac{1}{3}}, 1^{\frac{1}{1}}, 1^{\frac{1}{2}}, 1^{\frac{1}{2}} \right\}$. Similarly, the characteristic polynomial of the

$\{A, B, C\}$ matrices above is: $(\lambda^4 - 1) = 0$, and they have eigenvalues $\left\{ 1^{\frac{1}{4}}, 1^{\frac{1}{4}}, 1^{\frac{1}{4}}, 1^{\frac{1}{4}} \right\}$.

Since the set of powers of the pure permutation matrices are used to make the algebraic matrix forms, the $n \times n$ algebraic matrix forms will be associated with the nth roots of unity. We have seen this phenomenon in the non-matrix notation of the algebraic matrix forms.

However, these roots of unity will not be the euclidean complex number roots of unity. They are the roots of unity in algebras different from the euclidean complex numbers. In the case of the C_3 algebras, they are $\left\{ 1, \hat{p}, \hat{q} \right\}$. In the case of the C_4

algebras, they are $\left\{ 1, \hat{r}, \hat{s}, \hat{t} \right\}$. Even so, to every root of unity in an algebra other than the euclidean complex numbers, there corresponds a root of unity in the euclidean complex numbers, and so it is that the other algebras can be shadowed by the euclidean complex numbers.

Algebraically Isomorphism between the Algebraic Matrix Forms:

We have previously seen that the three algebraically isomorphic forms of the C_4 algebras differ from each other in that the element forming the C_2 sub-group is respectively $\left\{ C_{4.1} : c, C_{4.2} : b, C_{4.3} : d \right\}$. The reader might think that only groups with sub-groups have more than one (isomorphic) algebraic matrix form. This is not the case. The number of isomorphic forms depends upon the number of pure permutation matrices of a particular size, as we demonstrate below.

The 5-dimensional algebraic matrix form that is a copy of the standard form of the Cayley Table of C_5 is:

$$C_{5.1}L^1H^4{}_{(j=1,k=1,l=1,m=1)} = \begin{bmatrix} a & b & c & d & e \\ e & a & b & c & d \\ d & e & a & b & c \\ c & d & e & a & b \\ b & c & d & e & a \end{bmatrix}$$

If we raise a particular unit sub-matrix through integer powers, we get the other unit sub-matrices in a particular order. For example, the b unit sub-matrix of this algebra is:

$$B = \begin{bmatrix} 0 & 1 & 0 & 0 & 0 \\ 0 & 0 & 1 & 0 & 0 \\ 0 & 0 & 0 & 1 & 0 \\ 0 & 0 & 0 & 0 & 1 \\ 1 & 0 & 0 & 0 & 0 \end{bmatrix}$$

Raised to integer powers, this matrix becomes:

$$B^{1\ldots5} = B, C, D, E, A$$

There are other 5-dimensioal algebraic matrix forms. They are:

$$C_{5.2}L^1H^4 = \begin{bmatrix} a & b & c & d & e \\ c & a & d & e & b \\ b & e & a & c & d \\ e & d & b & a & c \\ d & c & e & b & a \end{bmatrix}, C_{5.3}L^1H^4 = \begin{bmatrix} a & b & c & d & e \\ c & a & e & b & d \\ b & d & a & e & c \\ e & c & d & a & b \\ d & e & b & c & a \end{bmatrix}$$

$$C_{5.4}L^1H^4 = \begin{bmatrix} a & b & c & d & e \\ d & a & e & c & b \\ e & c & a & b & d \\ b & e & d & a & c \\ c & d & b & e & a \end{bmatrix}$$

$$C_{5.5}L^1H^4 = \begin{bmatrix} a & b & c & d & e \\ e & a & d & b & c \\ d & c & a & e & b \\ c & e & b & a & d \\ b & d & e & c & a \end{bmatrix}, C_{5.6}L^1H^4 = \begin{bmatrix} a & b & c & d & e \\ d & a & b & e & c \\ e & d & a & c & b \\ b & c & e & a & d \\ c & e & d & b & a \end{bmatrix}$$

The respective b sub-matrix structures are:

$$L^!H^4{}_2 : B^{1...5} = B, E, D, C, A$$
$$L^!H^4{}_3 : B^{1...5} = B, D, E, C, A$$
$$L^!H^4{}_4 : B^{1...5} = B, E, C, D, A$$
$$L^!H^4{}_5 : B^{1...5} = B, D, C, E, A$$
$$L^!H^4{}_6 : B^{1...5} = B, C, E, D, A$$

So, even though the cyclic group of order five has no proper sub-groups, it does have six algebraically isomorphic algebraic matrix forms. There are $4! = 24$ 5×5 permutation matrices that have all zeros on the leading diagonal and which generate the group C_5 [73]. These are taken together in sets of four to make an algebraic matrix form. Hence six algebraically isomorphic algebraic matrix forms.

---

[73] There are other 5×5 permutation matrices with zeros on the leading diagonal. These matrices are effectively direct products of 2×2 and 3×3 permutation matrices.

Section 11 : Chapter 2

Permutation Matrices with Negative Entries

Although convention defines permutation matrices to have entries that are all unity, they can be written in smaller (half-size) matrices with entries that are unity or minus unity. The matrix:

$$B^1 = \begin{bmatrix} 0 & 0 & 1 & 0 \\ 0 & 0 & 0 & 1 \\ 0 & 1 & 0 & 0 \\ 1 & 0 & 0 & 0 \end{bmatrix}$$

is a permutation matrix that generates three other permutation matrices when raised to different powers. (It is a pure permutation matrix.)

$$B^2 = \begin{bmatrix} 0 & 1 & 0 & 0 \\ 1 & 0 & 0 & 0 \\ 0 & 0 & 0 & 1 \\ 0 & 0 & 1 & 0 \end{bmatrix}, B^3 = \begin{bmatrix} 0 & 0 & 0 & 1 \\ 0 & 0 & 1 & 0 \\ 1 & 0 & 0 & 0 \\ 0 & 1 & 0 & 0 \end{bmatrix}, B^4 = \begin{bmatrix} 1 & 0 & 0 & 0 \\ 0 & 1 & 0 & 0 \\ 0 & 0 & 1 & 0 \\ 0 & 0 & 0 & 1 \end{bmatrix}$$

The matrix:

$$C^1 = \begin{bmatrix} 0 & 1 \\ -1 & 0 \end{bmatrix}$$

also generates three matrices:

$$C^2 = \begin{bmatrix} -1 & 0 \\ 0 & -1 \end{bmatrix}, C^3 = \begin{bmatrix} 0 & -1 \\ 1 & 0 \end{bmatrix}, C^4 = \begin{bmatrix} 1 & 0 \\ 0 & 1 \end{bmatrix}$$

In both cases, the set of matrices forms the group C_4 under the operation of matrix multiplication. There is a one-to-one and onto mapping between these two sets of matrices.

$$B^1 : \begin{bmatrix} 0 & 0 & 1 & 0 \\ 0 & 0 & 0 & 1 \\ 0 & 1 & 0 & 0 \\ 1 & 0 & 0 & 0 \end{bmatrix} \rightarrow C^1 : \begin{bmatrix} 0 & 1 \\ -1 & 0 \end{bmatrix}, B^2 : \begin{bmatrix} 0 & 1 & 0 & 0 \\ 1 & 0 & 0 & 0 \\ 0 & 0 & 0 & 1 \\ 0 & 0 & 1 & 0 \end{bmatrix} \rightarrow C^2 : \begin{bmatrix} -1 & 0 \\ 0 & -1 \end{bmatrix}$$

$$B^3 : \begin{bmatrix} 0 & 0 & 0 & 1 \\ 0 & 0 & 1 & 0 \\ 1 & 0 & 0 & 0 \\ 0 & 1 & 0 & 0 \end{bmatrix} \rightarrow C^3 : \begin{bmatrix} 0 & -1 \\ 1 & 0 \end{bmatrix}, B^4 : \begin{bmatrix} 1 & 0 & 0 & 0 \\ 0 & 1 & 0 & 0 \\ 0 & 0 & 1 & 0 \\ 0 & 0 & 0 & 1 \end{bmatrix} \rightarrow C^4 : \begin{bmatrix} 1 & 0 \\ 0 & 1 \end{bmatrix}$$

These two sets of matrices are isomorphic over the operation of multiplication, as can be shown by multiplying together every pair within the set and comparing the result to the same product within the other set. Furthermore, by similar comparison of sums of pairs of matrices within a set, these two sets of matrices are isomorphic over the operation of addition. Equivalence under scalar multiplication is trivial.

> **Theorem:**
> The above two sets of matrices are algebraically equivalent.
> **Proof:**
> See above

Since algebraic matrix forms are constructed by turning a suitable set of permutation matrices into variable matrices and adding them. We have:

$$\begin{bmatrix} 0 & 0 & c & 0 \\ 0 & 0 & 0 & c \\ 0 & c & 0 & 0 \\ c & 0 & 0 & 0 \end{bmatrix} + \begin{bmatrix} 0 & b & 0 & 0 \\ b & 0 & 0 & 0 \\ 0 & 0 & 0 & b \\ 0 & 0 & b & 0 \end{bmatrix} + \begin{bmatrix} 0 & 0 & 0 & d \\ 0 & 0 & d & 0 \\ d & 0 & 0 & 0 \\ 0 & d & 0 & 0 \end{bmatrix} + \begin{bmatrix} a & 0 & 0 & 0 \\ 0 & a & 0 & 0 \\ 0 & 0 & a & 0 \\ 0 & 0 & 0 & a \end{bmatrix} = C_{4.2}L^1H^3_{(j=1,k=1,l=1)}$$

and:

$$\begin{bmatrix} 0 & c \\ -c & 0 \end{bmatrix} + \begin{bmatrix} -b & 0 \\ 0 & -b \end{bmatrix} + \begin{bmatrix} 0 & -d \\ d & 0 \end{bmatrix} + \begin{bmatrix} a & 0 \\ 0 & a \end{bmatrix} = \begin{bmatrix} a-b & c-d \\ -(c-d) & a-b \end{bmatrix} = \mathbb{C}(a-b,c-d)$$

This is, of course, the algebraic equivalence of the euclidean complex numbers and the $C_{4.2}L^1H^3_{(j=1,k=1,l=1)}$ algebra. The reader is urged to carefully consider the previous sentence.

$$\begin{bmatrix} a & b & c & d \\ b & a & d & c \\ d & c & a & b \\ c & d & b & a \end{bmatrix} \cong \begin{bmatrix} a-b & c-d \\ -(c-d) & a-b \end{bmatrix}$$

Of course, algebraic isomorphism does not mean that two algebras are identical in all ways.

> **Important Point:**
> The natural algebras are based upon permutation matrices. The $C_{4.2}L^1H^3_{(j=1,k=1,l=1)}$ algebra and the euclidean complex numbers are based upon equivalent sets of permutation matrices.

Introduction to Section 12

In this section, we introduce the reader to a new matrix operation, which we call folding and unfolding a matrix. This operation is not an algebraic operation like addition and multiplication. We contend that it is only a notation operation that changes from non-matrix notation to matrix notation.

However, when applied to the negative elements of algebraic matrix forms with negative space squeezing parameters, it unfolds the algebra into a higher dimensional algebra. This is quite startling, and it leads us to a very unconventional view of negative numbers. This view is contentious.

Section 12 : Chapter 1

The Folding Operation

It is a basic property of matrices that they can be partitioned into blocks and multiplied together as if the blocks were the elements of a matrix. When this is done, the result is the same as if the matrices were not partitioned. We give an example:

$$
\begin{bmatrix} a & b & c & d \\ e & f & g & h \\ i & j & k & l \\ m & n & o & p \end{bmatrix}
\begin{bmatrix} \alpha & \beta & \chi & \delta \\ \varepsilon & \phi & \varphi & \gamma \\ \eta & \iota & \kappa & \lambda \\ \mu & \nu & o & \pi \end{bmatrix} =
$$

$$
\begin{bmatrix}
a\alpha + b\varepsilon + c\eta + d\mu & a\beta + b\phi + c\iota + d\nu & a\chi + b\varphi + c\kappa + do & a\delta + b\lambda + c\lambda + d\pi \\
e\alpha + f\varepsilon + g\eta + h\mu & e\beta + f\phi + g\iota + h\nu & e\chi + f\varphi + g\kappa + ho & e\delta + f\lambda + g\lambda + h\pi \\
i\alpha + j\varepsilon + k\eta + l\mu & i\beta + j\phi + k\iota + l\nu & i\chi + j\varphi + k\kappa + lo & i\delta + j\lambda + k\lambda + l\pi \\
m\alpha + n\varepsilon + o\eta + p\mu & m\beta + n\phi + o\iota + p\nu & m\chi + n\varphi + o\kappa + po & m\delta + n\lambda + o\lambda + p\pi
\end{bmatrix}
$$

$$
\begin{bmatrix}
\begin{bmatrix} a & b \\ e & f \end{bmatrix} & \begin{bmatrix} c & d \\ g & h \end{bmatrix} \\
\begin{bmatrix} i & j \\ m & n \end{bmatrix} & \begin{bmatrix} k & l \\ o & p \end{bmatrix}
\end{bmatrix}
\begin{bmatrix}
\begin{bmatrix} \alpha & \beta \\ \varepsilon & \phi \end{bmatrix} & \begin{bmatrix} \chi & \delta \\ \varphi & \lambda \end{bmatrix} \\
\begin{bmatrix} \eta & \iota \\ \mu & \nu \end{bmatrix} & \begin{bmatrix} \kappa & \lambda \\ o & \pi \end{bmatrix}
\end{bmatrix} =
$$

$$
\begin{bmatrix}
\begin{bmatrix} a & b \\ e & f \end{bmatrix}\begin{bmatrix} \alpha & \beta \\ \varepsilon & \phi \end{bmatrix} + \begin{bmatrix} c & d \\ g & h \end{bmatrix}\begin{bmatrix} \eta & \iota \\ \mu & \nu \end{bmatrix} & \begin{bmatrix} a & b \\ e & f \end{bmatrix}\begin{bmatrix} \chi & \delta \\ \varphi & \lambda \end{bmatrix} + \begin{bmatrix} c & d \\ g & h \end{bmatrix}\begin{bmatrix} \kappa & \lambda \\ o & \pi \end{bmatrix} \\
\begin{bmatrix} i & j \\ m & n \end{bmatrix}\begin{bmatrix} \alpha & \beta \\ \varepsilon & \phi \end{bmatrix} + \begin{bmatrix} k & l \\ o & p \end{bmatrix}\begin{bmatrix} \eta & \iota \\ \mu & \nu \end{bmatrix} & \begin{bmatrix} i & j \\ m & n \end{bmatrix}\begin{bmatrix} \chi & \delta \\ \varphi & \lambda \end{bmatrix} + \begin{bmatrix} k & l \\ o & p \end{bmatrix}\begin{bmatrix} \kappa & \lambda \\ o & \pi \end{bmatrix}
\end{bmatrix} =
$$

$$
\begin{bmatrix}
\begin{bmatrix} a\alpha + b\varepsilon + c\eta + d\mu & a\beta + b\phi + c\iota + d\nu \\ e\alpha + f\varepsilon + g\eta + h\mu & e\beta + f\phi + g\iota + h\nu \end{bmatrix} & \begin{bmatrix} a\chi + b\varphi + c\kappa + do & a\delta + b\lambda + c\lambda + d\pi \\ e\chi + f\varphi + g\kappa + ho & e\delta + f\lambda + g\lambda + h\pi \end{bmatrix} \\
\begin{bmatrix} i\alpha + j\varepsilon + k\eta + l\mu & i\beta + j\phi + k\iota + l\nu \\ m\alpha + n\varepsilon + o\eta + p\mu & m\beta + n\phi + o\iota + p\nu \end{bmatrix} & \begin{bmatrix} i\chi + j\varphi + k\kappa + lo & i\delta + j\lambda + k\lambda + l\pi \\ m\chi + n\varphi + o\kappa + po & m\delta + n\lambda + o\lambda + p\pi \end{bmatrix}
\end{bmatrix}
$$

Block addition obviously gives the same results as normal addition. Thus, we have covered both the algebraic operations.

The matrix representation of euclidean complex numbers is:

$$a + i\hat{b} \equiv \begin{bmatrix} a & b \\ -b & a \end{bmatrix}$$

Consider a matrix with euclidean complex numbers as elements; we can replace each euclidean complex number with a 2×2 matrix of the complex number form.

$$\begin{bmatrix} a+i\hat{b} & c+i\hat{d} \\ e+i\hat{f} & g-i\hat{h} \end{bmatrix} \triangleleft \begin{bmatrix} \begin{bmatrix} a & b \\ -b & a \end{bmatrix} & \begin{bmatrix} c & d \\ -d & c \end{bmatrix} \\ \begin{bmatrix} e & f \\ -f & e \end{bmatrix} & \begin{bmatrix} g & -h \\ h & g \end{bmatrix} \end{bmatrix} = \begin{bmatrix} a & b & c & d \\ -b & a & -d & c \\ e & f & g & -h \\ -f & e & h & g \end{bmatrix}$$

We refer to such expansion from a smaller matrix to a larger matrix as unfolding the matrix. We refer to the inverse operation as folding the matrix.

Notation:

$$\{\triangleleft, \triangleright\}$$

Reading left to right, we use the $\triangleleft$ sign to indicate unfolding a smaller matrix into a larger matrix. Reading left to right, we use the $\triangleright$ sign to indicate folding a larger matrix into a smaller matrix.

The algebraic equivalence of normal matrices and blocked matrices guarantees that the large matrix is algebraically the same as the small matrix with the complex elements. We have effectively 'unfolded' the complex numbers into real numbers. The procedure is reversible if the 4×4 matrices contain blocks of the form of a complex number.

Folding and unfolding an algebraic matrix form in this way is no more than changing the notation. We have been doing this with complex numbers from the start of this work:

$$\begin{bmatrix} a+i\hat{b} \end{bmatrix} \triangleleft \begin{bmatrix} a & b \\ -b & a \end{bmatrix}$$

We can unfold a single element into an overblown representation using the algebraic isomorphism of the real number matrices.

$$[a] \triangleleft \begin{bmatrix} a & 0 \\ 0 & a \end{bmatrix}$$

By this means, we obtain an overblown representation of an algebra.

$$\left[a+\hat{i}b\right] \lhd \begin{bmatrix} a & b \\ -b & a \end{bmatrix} \lhd \begin{bmatrix} a & 0 & b & 0 \\ 0 & a & 0 & b \\ -b & 0 & a & 0 \\ 0 & -b & 0 & a \end{bmatrix}$$

Again, algebraically, this is a notational change, but not all things (square roots) are algebraic operations.

We are not restricted to euclidean complex numbers. Because the 4-dimensional $C_{4.2}$ algebraic matrix form is four (blocks of) 2-dimensional numbers, it can be folded:

$$C_{4.2}L^lE^3_{(j=-1,k=-1,l=1)}: \begin{bmatrix} a & b & c & d \\ -b & a & -d & c \\ d & -c & a & b \\ c & d & -b & a \end{bmatrix} \rhd \begin{bmatrix} a+\hat{i}b & c+\hat{i}d \\ d-\hat{i}c & a+\hat{i}b \end{bmatrix} \rhd a+\hat{s}b+\hat{t}c+\hat{u}d$$

$$C_{4.2}L^lE^3_{(j=-1,k=-1,l=-1)}: \begin{bmatrix} a & b & c & d \\ -b & a & -d & c \\ -d & c & a & b \\ -c & -d & -b & a \end{bmatrix} \rhd \begin{bmatrix} a+\hat{i}b & c+\hat{i}d \\ -d+\hat{i}c & a+\hat{i}b \end{bmatrix} \rhd a+\hat{s}b+\hat{t}c+\hat{u}d$$

$$C_{4.2}L^lH^3_{(j=1,k=1,l=-1)}: \begin{bmatrix} a & b & c & d \\ b & a & d & c \\ -d & -c & a & b \\ -c & -d & b & a \end{bmatrix} \rhd \begin{bmatrix} a+\hat{r}b & c+\hat{r}d \\ -d-\hat{r}c & a+\hat{r}b \end{bmatrix} \rhd a+\hat{s}b+\hat{t}c+\hat{u}d$$

$$C_{4.2}L^lH^3_{(j=1,k=1,l=1)}: \begin{bmatrix} a & b & c & d \\ b & a & d & c \\ d & c & a & b \\ c & d & b & a \end{bmatrix} \rhd \begin{bmatrix} a+\hat{r}b & c+\hat{r}d \\ d+\hat{r}c & a+\hat{r}b \end{bmatrix} \rhd a+\hat{s}b+\hat{t}c+\hat{u}d$$

where $\hat{r}=\sqrt{+1}$.

162

The $C_{4.2}L^1E^3{}_{(j=-1,k=-1,l=1)}$ algebra is:

$$\begin{bmatrix} a & b & c & d \\ -b & a & -d & c \\ d & -c & a & b \\ c & d & -b & a \end{bmatrix}$$

We have:

$$\left\{ \begin{bmatrix} a & 0 & 0 & 0 \\ 0 & a & 0 & 0 \\ 0 & 0 & a & 0 \\ 0 & 0 & 0 & a \end{bmatrix} \triangleright \begin{bmatrix} a & 0 \\ 0 & a \end{bmatrix}, \begin{bmatrix} 0 & b & 0 & 0 \\ -b & 0 & 0 & 0 \\ 0 & 0 & 0 & b \\ 0 & 0 & -b & 0 \end{bmatrix} \triangleright \begin{bmatrix} \hat{i}b & 0 \\ 0 & \hat{i}b \end{bmatrix}, \right.$$

$$\left. \begin{bmatrix} 0 & 0 & c & 0 \\ 0 & 0 & 0 & c \\ 0 & -c & 0 & 0 \\ c & 0 & 0 & 0 \end{bmatrix} \triangleright \begin{bmatrix} 0 & c \\ -\hat{i}c & 0 \end{bmatrix}, \begin{bmatrix} 0 & 0 & 0 & d \\ 0 & 0 & -d & 0 \\ d & 0 & 0 & 0 \\ 0 & d & 0 & 0 \end{bmatrix} \triangleright \begin{bmatrix} 0 & \hat{i}d \\ d & 0 \end{bmatrix} \right\}$$

Adding the four 2×2 matrices together gives the 2×2 form of the $C_{4.2}L^1E^3{}_{(j=-1,k=-1,l=1)}$ algebra:

$$\begin{bmatrix} a+\hat{i}b & c+\hat{i}d \\ d-\hat{i}c & a+\hat{i}b \end{bmatrix}$$

In non-matrix notation, the $C_{4.2}L^1E^3{}_{(j=-1,k=-1,l=1)}$ algebra is: $a+\hat{s}b+\hat{t}c+\hat{u}d$. Thus:

$$a+\hat{s}b+\hat{t}c+\hat{u}d \triangleleft \begin{bmatrix} a+\hat{i}b & c+\hat{i}d \\ d-\hat{i}c & a+\hat{i}b \end{bmatrix}$$

Section 12 : Chapter 2

Unfolding Negative Real Numbers

The negative real numbers are not an algebra because they are not multiplicatively closed and they do not include the multiplicative identity. Matrices of the form:

$$\begin{bmatrix} 0 & b \\ b & 0 \end{bmatrix}$$

have exactly the same algebraic failings. Furthermore, over the algebraic operations, these matrices are isomorphic to the negative real numbers.

Theorem:

Matrices of the form $\begin{bmatrix} 0 & b \\ b & 0 \end{bmatrix}$ are isomorphic to the negative real numbers over the algebraic operations.

Proof:

$$\phi:(-\mathbb{R},+,\times) \mapsto \left(M_{(2,\mathbb{R})},+,\bullet\right) \quad : \quad \phi([-b]) \mapsto \begin{bmatrix} 0 & b \\ b & 0 \end{bmatrix}$$

i) $\quad \phi([-c]+[-d]) \mapsto \begin{bmatrix} 0 & c \\ c & 0 \end{bmatrix} + \begin{bmatrix} 0 & d \\ d & 0 \end{bmatrix} = \begin{bmatrix} 0 & c+d \\ c+d & 0 \end{bmatrix}$

ii) $\quad \phi([-c]\times[-d]) \mapsto \begin{bmatrix} 0 & c \\ c & 0 \end{bmatrix} \bullet \begin{bmatrix} 0 & d \\ d & 0 \end{bmatrix} = \begin{bmatrix} cd & 0 \\ 0 & cd \end{bmatrix}$

iii) $\quad \phi(\alpha[-c]) \mapsto \begin{bmatrix} \alpha & 0 \\ 0 & \alpha \end{bmatrix}\begin{bmatrix} 0 & c \\ c & 0 \end{bmatrix} = \begin{bmatrix} 0 & \alpha c \\ \alpha c & 0 \end{bmatrix}$

Because these matrices are isomorphic to the negative real numbers in all algebraic matters, we can unfold negative real numbers into this matrix form, and we can fold such matrices into negative numbers. Within an algebraic matrix form with negative space squeezing parameters, we can unfold the negative space squeezing parameters. For example, we can unfold the euclidean complex numbers:

$$\begin{bmatrix} a & b \\ -b & a \end{bmatrix} \triangleleft \begin{bmatrix} a & 0 & b & 0 \\ 0 & a & 0 & b \\ 0 & b & a & 0 \\ b & 0 & 0 & a \end{bmatrix}$$

When we unfold an algebraic matrix form with negative space squeezing parameters, the larger matrix might not be multiplicatively closed because the block multiplication property of matrices might bring about a change of form from the 'negative number blocks'.

Digression:

The hyperbolic complex numbers folded

Matrices of the appropriate form can be folded into a negative real number.

$$\begin{bmatrix} 0 & b \\ b & 0 \end{bmatrix} \triangleright -b$$

More than this, we can fold matrices of the form:

$$\begin{bmatrix} a & b \\ b & a \end{bmatrix} = \left(\begin{bmatrix} a & 0 \\ 0 & a \end{bmatrix} + \begin{bmatrix} 0 & b \\ b & 0 \end{bmatrix} \right) \triangleright a - b$$

$$a + \hat{r}b \equiv \begin{bmatrix} a & b \\ b & a \end{bmatrix} \triangleright a - b$$

This is consistent with $\hat{r} = \sqrt{+1}$ if we take $\hat{r}$ to be minus unity. However, if we reduce the $a-b$ object to a single real number, we lose a dimension - two dimensions appear to fold into one. Within the 2×2 matrix, it makes no sense to subtract one element from another.

Section 12 : Chapter 3

A Chain of Algebras

The $C_{4.2}L^1H^3{}_{(j=1,k=1,l=1)}$ algebra is not the only 4-dimensional algebra. For example, by setting the space squeezing parameters as $\{j=-1,k=-1,l=1\}$, we get:

$$C_{4.2}L^1E^3{}_{(j=-1,k=-1,l=1)} : \begin{bmatrix} a & c & b & d \\ -c & a & -d & b \\ d & -b & a & c \\ b & d & -c & a \end{bmatrix}$$

This algebra can be folded:

$$\begin{bmatrix} a & c & b & d \\ -c & a & -d & b \\ d & -b & a & c \\ b & d & -c & a \end{bmatrix} \triangleright \begin{bmatrix} a+\hat{i}b & c+\hat{i}d \\ d-\hat{i}c & a+\hat{i}b \end{bmatrix} \triangleright a+b\hat{s}+c\hat{t}+d\hat{u}$$

It can be unfolded:

$$\begin{bmatrix} a & c & b & d \\ -c & a & -d & b \\ d & -b & a & c \\ b & d & -c & a \end{bmatrix} \triangleleft \begin{bmatrix} a & 0 & c & 0 & b & 0 & d & 0 \\ 0 & a & 0 & c & 0 & b & 0 & d \\ 0 & c & a & 0 & 0 & d & b & 0 \\ c & 0 & 0 & a & d & 0 & 0 & b \\ d & 0 & 0 & b & a & 0 & c & 0 \\ 0 & d & b & 0 & 0 & a & 0 & c \\ b & 0 & d & 0 & 0 & c & a & 0 \\ 0 & b & 0 & d & c & 0 & 0 & a \end{bmatrix}$$

This is the $C_8L^1H^7{}_{(j=1,...,p=1)}$ algebra (with some variables equal to zero).

Thus, we have a chain of algebras. We start with the 2-dimensional hyperbolic complex numbers, and we introduce a negative space squeezing parameter thereby forming the euclidean complex numbers. We unfold the euclidean complex numbers into the $C_{4.2}L^1H^3{}_{(j=1,k=1,l=1)}$ algebra. We introduce a set of negative space squeezing parameters and unfold the matrix and get the $C_8L^1H^7{}_{(j=1,...,p=1)}$ algebra.

Of course, we cannot unfold a 2-dimensiional algebra into a 3-dimensional algebra. The algebras must double in size at each unfolding.

Section 12 : Chapter 4

Unfolding the Euclidean Complex Numbers

In a previous chapter, we unfolded the euclidean complex numbers and thereby removed the negative space squeezing parameter. The unfolded form of the euclidean complex numbers is the matrix:

$$\begin{bmatrix} a & c \\ -c & a \end{bmatrix} \vartriangleleft \begin{bmatrix} a & 0 & c & 0 \\ 0 & a & 0 & c \\ 0 & c & a & 0 \\ c & 0 & 0 & a \end{bmatrix}$$

Matrices of this form are not multiplicatively closed. They expand under multiplication to become the 4-dimensional $C_{4.2}L^{l}H^{3}{}_{(j=1,k=1,l=1)}$ algebra.

$$\begin{bmatrix} a & 0 & c & 0 \\ 0 & a & 0 & c \\ 0 & c & a & 0 \\ c & 0 & 0 & a \end{bmatrix} \begin{bmatrix} e & 0 & g & 0 \\ 0 & e & 0 & g \\ 0 & g & e & 0 \\ g & 0 & 0 & e \end{bmatrix} = \begin{bmatrix} ae & cg & ag+ce & 0 \\ cg & ae & 0 & ag+ce \\ 0 & ag+ce & ae & cg \\ ag+ce & 0 & cg & ae \end{bmatrix}$$

$$\left(a+\hat{i}c\right)\left(e+\hat{i}g\right)=\left(ae-cg\right)+\hat{i}\left(ag+ce\right)$$

$$\begin{bmatrix} l & 0 & m & 0 \\ 0 & l & 0 & m \\ 0 & m & l & 0 \\ m & 0 & 0 & l \end{bmatrix} \begin{bmatrix} ae & cg & ag+ce & 0 \\ cg & ae & 0 & ag+ce \\ 0 & ag+ce & ae & cg \\ ag+ce & 0 & cg & ae \end{bmatrix}$$

$$= \begin{bmatrix} ael & cgl+agm+cem & agl+cel+aem & cgm \\ cgl+agm+cem & ael & cgm & agl+cel+aem \\ cgm & agl+cel+aem & ael & cgl+agm+cem \\ agl+cel+aem & cgm & cgl+agm+cem & ael \end{bmatrix}$$

Folding this gives:

$$\begin{bmatrix} ael & cgl+agm+cem & agl+cel+aem & cgm \\ cgl+agm+cem & ael & cgm & agl+cel+aem \\ cgm & agl+cel+aem & ael & cgl+agm+cem \\ agl+cel+aem & cgm & cgl+agm+cem & ael \end{bmatrix}$$

$$\vartriangleright \begin{bmatrix} ael-cgl-agm-cem & agl+cel+aem-cgm \\ -\left(agl+cel+aem-cgm\right) & ael-cgl-agm-cem \end{bmatrix}$$

We have:

$$\left(a+\hat{i}c\right)\left(e+\hat{i}g\right)\left(l+\hat{i}m\right)=\left(ael-cgl-agm-cem\right)+\hat{i}\left(agl+cel+aem-cgm\right)$$

The $C_{4.2}L^{1}H^{3}{}_{(j=1,k=1,l=1)}$ algebra is, of course, closed under multiplication.

The polar form of the $C_{4.2}L^{1}H^{3}{}_{(j=1,k=1,l=1)}$ algebra is:

$$\exp\left(\begin{bmatrix} a & b & c & d \\ b & a & d & c \\ d & c & a & b \\ c & d & b & a \end{bmatrix}\right) = PROD\left(\begin{bmatrix} h & 0 & 0 & 0 \\ 0 & h & 0 & 0 \\ 0 & 0 & h & 0 \\ 0 & 0 & 0 & h \end{bmatrix}\begin{bmatrix} \cosh(b) & \sinh(b) & 0 & 0 \\ \sinh(b) & \cosh(b) & 0 & 0 \\ 0 & 0 & \cosh(b) & \sinh(b) \\ 0 & 0 & \sinh(b) & \cosh(b) \end{bmatrix}\right.$$

$$\begin{bmatrix} AH_4(c) & CH_4(c) & BH_4(c) & DH_4(c) \\ CH_4(c) & AH_4(c) & DH_4(c) & BH_4(c) \\ DH_4(c) & BH_4(c) & AH_4(c) & CH_4(c) \\ BH_4(c) & DH_4(c) & CH_4(c) & AH_4(c) \end{bmatrix}$$

$$\left.\begin{bmatrix} AH_4(d) & CH_4(d) & DH_4(d) & BH_4(d) \\ CH_4(d) & AH_4(d) & BH_4(d) & DH_4(d) \\ BH_4(d) & DH_4(d) & AH_4(d) & CH_4(d) \\ DH_4(d) & BH_4(d) & CH_4(d) & AH_4(d) \end{bmatrix}\right)$$

We fold this into 2×2 matrices.

$$PROD\left(\begin{bmatrix} e^a & 0 \\ 0 & e^a \end{bmatrix}\begin{bmatrix} \cosh(b)-\sinh(b) & 0 \\ 0 & \cosh(b)-\sinh(b) \end{bmatrix}\right.$$

$$\begin{bmatrix} AH_4(c)-CH_4(c) & BH_4(c)-DH_4(c) \\ DH_4(c)-BH_4(c) & AH_4(c)-CH_4(c) \end{bmatrix}$$

$$\left.\begin{bmatrix} AH_4(d)-CH_4(d) & DH_4(d)-BH_4(d) \\ BH_4(d)-DH_4(d) & AH_4(d)-CH_4(d) \end{bmatrix}\right)$$

$$=\begin{bmatrix} e^a & 0 \\ 0 & e^a \end{bmatrix}\begin{bmatrix} e^{-b} & 0 \\ 0 & e^{-b} \end{bmatrix}\begin{bmatrix} \cos(c) & \sin(c) \\ -\sin(c) & \cos(c) \end{bmatrix}\begin{bmatrix} \cos(d) & -\sin(d) \\ \sin(d) & \cos(d) \end{bmatrix}$$

$$=\begin{bmatrix} e^{a-b} & 0 \\ 0 & e^{a-b} \end{bmatrix}\begin{bmatrix} \cos(c-d) & \sin(c-d) \\ -\sin(c-d) & \cos(c-d) \end{bmatrix}$$

Two rotation matrices have combined into a single rotation matrix, and one has turned into a distance.
Depolarizing it gives:

$$\begin{bmatrix} a-b & c-d \\ -(c-d) & a-b \end{bmatrix}$$

We could have obtained this directly by folding the $C_{4.2}L^{1}H^{3}{}_{(j=1,k=1,l=1)}$ algebra:

$$\begin{bmatrix} a & b & c & d \\ b & a & d & c \\ d & c & a & b \\ c & d & b & a \end{bmatrix} \triangleright \begin{bmatrix} a-b & c-d \\ -(c-d) & a-b \end{bmatrix}$$

The $C_{4.2}L^1H^3{}_{(j=1,k=1,l=1)}$ algebra contains a 2-dimensional sub-algebra. This sub-algebra is the hyperbolic complex numbers (with overblown representation):

$$\begin{bmatrix} a & b & 0 & 0 \\ b & a & 0 & 0 \\ 0 & 0 & a & b \\ 0 & 0 & b & a \end{bmatrix}$$

The euclidean complex number algebra is a folded form of the $C_{4.2}L^1H^3{}_{(j=1,k=1,l=1)}$ algebra. However, the folded form of an algebra is not the same thing as the unfolded form of that algebra. The axes of the unfolded algebra point in different directions; this is not so in the folded form. The folding has arranged the axes in a particular way – it has paired them. The sub-algebra is not apparent, and the distance functions are changed.

The euclidean complex numbers contain the 2-dimensional euclidean space. The $C_{4.2}L^1H^3{}_{(j=1,k=1,l=1)}$ algebra contains 4-dimensional space.

That the euclidean complex numbers are a folded form of a 4-dimensional algebra is consistent with the fact that the unit sub-matrices of the euclidean complex numbers form the group C_4.

Theorem:

The euclidean complex numbers are a folded form of the 4-dimensional C_4 algebra.

Proof:

See above.

Section 12 : Chapter 5

Unfolding in General

In Section 5, we discovered that any $n\times n$ matrix that has n variables arranged so that each variable appears once and only once in every row and in every column and has the same variable as every element on the leading diagonal is an algebraic matrix form if it is multiplicatively commutative. This matrix form is guaranteed non-singular in its polar form. Clearly, any unfolded algebraic matrix form satisfies these requirements (with half of the variables at zero) if it is multiplicatively commutative and it is not a sub-algebra of a higher dimensioned algebra. (If the unfolded algebraic matrix form is a sub-algebra of the larger algebraic matrix form, it will not multiplicatively expand into the larger algebra.)

We are unfolding algebraic matrix forms with negative space squeezing parameters. Any unfolded algebraic matrix form can be arranged as 2×2 blocks, one for each of the elements of the folded form. Because there is at least one negative space squeezing parameter, at least one of these blocks will be of the matrix form of a negative number. Such a block will multiply to form a real number matrix. The block multiplication property of matrices ensures that this property will be a property of the larger matrix. Thus it is that an unfolded algebraic matrix form cannot be a sub-algebra of the larger matrix.

Each 2×2 block of the unfolded algebraic matrix form is of the form of either:

$$\left\{ \begin{bmatrix} a & 0 \\ 0 & a \end{bmatrix}, \begin{bmatrix} 0 & b \\ b & 0 \end{bmatrix} \right\}$$

Both these matrix forms are multiplicatively commutative with each other and with themselves. Since the algebraic matrix forms with the negative space squeezing parameters are multiplicatively commutative, and the blocks that we have unfolded the elements of that matrix into are multiplicatively commutative, the unfolded algebraic matrix form will be multiplicatively commutative.

Every time we unfold an algebraic matrix form with negative space squeezing parameters, we will get an algebraic matrix form that has no negative space squeezing parameters. There is only one such algebra (usually with many algebraically isomorphic matrix forms). Thus, every one of the algebraic matrix forms with negative space squeezing parameters of a particular size will unfold into the same algebra. In the case of the three 3-dimensional algebraic matrix forms with negative space squeezing parameters, they all unfold into the $C_6 L^1 H^5_{(j=1,...,n=1)}$ algebra. All the 4-dimensional algebraic matrix forms with negative space squeezing parameters unfold into the $C_8 L^1 H^7_{(j=1,...,p=1)}$ algebra.

So it is that the n-dimensional algebraic matrix forms with negative space squeezing parameters are folded forms of the $2n$-dimensional algebraic matrix forms with all positive space squeezing parameters.

When we fold a $2n$-dimensional algebraic matrix form into a n-dimensional algebraic matrix form, the elements of the n-dimensional algebraic matrix form are of the form: $(a-c)$. This mirrors the simple trigonometric functions where we have:

$$\cos(\theta) = AH_4(\theta) - CH_4(\theta).$$

The n-dimensional algebraic matrix forms with negative space squeezing parameters have unit sub-matrices that form the group C_{2n}. This is consistent with them being considered to be $2n$-dimensional algebras.

Thus, we effectively have one natural algebra for each cyclic group. The cross products of these groups have cross-product algebras associated with them.

Theorem:
　　There is effectively one and only one natural algebra for every cyclic group.
Proof:
　　See above.

The natural algebra associated with a particular cyclic group has a geometric space within it. This space is always an H-type space with H-type trigonometric functions because the natural algebra has no negative space squeezing parameters.

A Simplification:
　　The above theorem is a substantial simplification of the natural algebras. We now have it that a number is a cyclic group is an algebra is a geometric space is a algebraic vector product space, and that there is one, and only one, such number of order n. Of course, we have to accept the concept of a folded algebra to achieve this simplification. With the folded algebra comes the folded space.

Section 12 : Chapter 6

A Farewell to Negative Numbers

In non-matrix notation, the euclidean complex numbers include the imaginary number $\hat{i}$. Clifford algebraists see this entity as a bivector; most mathematicians see it as a number of equal status to any real number. However, when we write the euclidean complex numbers in matrix form:

$$\begin{bmatrix} a & b \\ -b & a \end{bmatrix}$$

we see that it is nothing more than a notational device for keeping track of the matrix configuration:

$$\begin{bmatrix} 0 & 1 \\ -1 & 0 \end{bmatrix}$$

Similarly, when we unfold a matrix containing negative real numbers, we see that the minus sign is nothing more than a notational device for keeping track of the matrix configuration:

$$\begin{bmatrix} 0 & 1 \\ 1 & 0 \end{bmatrix}$$

There stands the argument; let the reader decide its merit.

Theorem (contentious):
Negative numbers are no more than a notational device.
Proof:
See above

The above theorem does not prevent us from doing subtraction of real numbers provided the result is positive or zero. We knew this when we were at infant school where we learnt things like "three take-away five – can't do". It is only education that has led us to think that we can do such sums.

Within the physical universe, nothing exists in negative quantities[74], and so we ought to be glad to be rid of negative numbers.

Without negative numbers, a spatial axis runs from zero to plus infinity rather than from minus infinity to plus infinity. Thus it is that the '2-dimensional' euclidean plane is 4-dimensional with axes aligned in pairs.

The axioms of an algebraic field include an axiom requiring the inclusion of additive inverses. This axiom is independent of all other axioms (that is why it is an axiom), and it can be removed without upsetting anything[75]. If we do this, we still have an intact algebraic structure that we might refer to as a positive definite algebraic field.

---

[74] Electrons have a positive amount of negative electric charge.
[75] Except every mathematician on planet Earth.

In a positive definite algebraic field, we still have subtraction, but we view it as a form of addition – subtracting three from five expresses the same relation as adding three and two. Of course, we cannot do subtraction if the result is less than zero.

Concluding Remarks

Nomenclature:

Spaces with Quadratic Signature

A space with a quadratic signature is a geometric space with a distance function that is the square root of a quadratic expression. For example:

$$\left\{ d = \sqrt{x^2 + y^2 + z^2 + t^2}, d = \sqrt{t^2 - y^2} \right\}$$

The existence of the natural spaces compels us to radically rethink our concepts of space. We now see that we live in two 2-dimensional spaces rather than one 4-dimensional space and that one of these 2-dimensional spaces is the folded form of a 4-dimensional space from which the other 2-dimensional sub-space has been extracted.

In the 1-dimensional case, the natural space is the same in all respects as the 1-dimensional space with which we are familiar. In the 2-dimensional case, the natural spaces are the same in all respects as the 2-dimensional euclidean space and 2-dimensional hyperbolic space with which we are familiar. There is, however, a difference of viewpoint between the 2-dimensional natural spaces and the spaces with quadratic signature. Mathematicians are accustomed to thinking of 2-dimensional euclidean space as being two 1-dimensional spaces set at right angles to each other. The natural 2-dimensional euclidean space is intrinsically 2-dimensional; it has only one 1-dimensional space (the real axis); the other axis (the imaginary one) is not a 1-dimensional space - $1 \times 1 = 1$, but $\hat{i} \times \hat{i} \neq \hat{i}$.

In general, mathematicians view *n*-dimensional space with quadratic signature as a construct of *n* 1-dimensional spaces fixed together at right angles to each other. The natural spaces are different in that the *n*-dimensional natural spaces have only one 1-dimensional space (the real axis) and all the other axes are similar to the imaginary axis in the euclidean complex numbers.

Since 1-dimensional space does not have angles in it[76], it is untenable to fix 1-dimensional spaces together at right angles to each other. In practice, mathematicians use the 2-dimensional euclidean trigonometric functions to fix 1-dimensional spaces together to form the spaces with quadratic signature.

Spaces with quadratic signature can have one or more 1-dimensional spaces stripped from them to form a lesser dimensional space. This can only happen in the natural spaces if the particular natural space contains a sub-space of the appropriate dimension to which it can be stripped back. This is another aspect of the intrinsic *n*-dimensionality of the natural spaces.

An angle in 2-dimensional hyperbolic space is a very different thing from an angle in 2-dimensional euclidean space. So it is with all the natural algebras; each one has its own type of angle. This contrasts sharply with the concept of angle used in spaces with quadratic signature. Each of the natural spaces has its own set of trigonometric

---

[76] The 1-dimensional trigonometric function, unity, does not accept an argument.

174

functions whereas the spaces with quadratic signature always use the 2-dimensional trigonometric functions.

Rotations within the natural spaces are all commutative. This is also the case in the 2-dimensional spaces with quadratic signature, but, in the higher dimensional spaces with quadratic signature, the rotations do not commute.

The natural spaces are intimately connected to abelian groups; indeed, it can be said that they are the abelian groups. There is no such intimate group connection in the spaces with quadratic signature.

One outstanding property of the natural spaces is that vectors can be formulated as an algebraic field within a natural space, and the natural spaces are the only spaces within which this can be done. Within the spaces with quadratic signature, 'vector algebra' is not an algebra but just a linear space with two weird 'multiplication' operations. If the universe is mathematical, then a universe based upon the natural spaces is much tidier than one based upon the spaces with quadratic signature.

However, the most impressive thing about the natural spaces is that they are discovered not invented. The particular distance functions, inner products, rotation matrices, and trigonometric functions are forced upon the mathematician by the natural algebra of the natural space. This is in profound contradistinction to the spaces with quadratic signature where the mathematician imposes the distance function and inner product upon the space. The existence of the natural spaces is based upon the existence of the real numbers alone; we need no more postulates than that to form the natural spaces. It is the distributive property of the real numbers that gives rise to the concept of linear transformation, and, with that, the natural algebras and the natural spaces fall, fully constructed, at our feet.

The hyperbolic complex numbers are restricted in cartesian form. That restriction is: $a \geq \phi b : \phi = 1$. Since we know that 2-dimensional hyperbolic space manifests itself to us as space-time, we know that the parameter ϕ is related to the speed of light. Thus, from no more than the existence of the real numbers, we have a physical constant falling out of the algebra.

We perceive the two 2-dimensional natural spaces as geometric spaces. Since we do not perceive the higher dimensional spaces as geometric spaces, it seems that they must manifest themselves in a non-geometrical way. If we attempt (and it is wrong to do this) to take the clifford algebra of the 3-dimensional spaces, the tri-vectors appear in three singlets, three pairs, and one trio; the sub-atomic particles come to mind.

Appendix 1

Axioms of a Metric

A function of the form: distance $= f(a,b,...)$ can be used as a metric if:

i) The distance from a point to itself is zero: $d(a,a) = 0$

ii) The distance from a to b is the same as the distance from b to a: $d(a,b) = d(b,a)$

iii) The distance from a to b via c is greater than or equal to the distance from a to b (think of a triangle): $d(a,b) \leq d(a,c) + d(c,b)$ with equality only if c lies on the straight line between a and b.

The second of these axioms assumes one can travel both from a to b and from b to a. Although this is possible in 2-dimensional euclidean space, it might not be possible in a 2-dimensional space-time.

The third of these is violated by Minkowski space-time which is the space of special relativity and hence of reality. Minkowski space-time has the distance function: $x^2 - y^2 = h^2$ where h is the hypotenuse of the triangle. Clearly, the length of the hypotenuse is less than the length of the x side.

Axioms of an Inner Product

An inner product of two vectors is denoted $\langle \underline{a}, \underline{b} \rangle$. An example is the dot product:

$$\langle \underline{a}, \underline{b} \rangle = \underline{a} \bullet \underline{b} = a_1 b_1 + a_2 b_2 + a_3 b_3$$

The inner product of two vectors is a projection function that produces the length of one vector projected on to the other. To qualify as an inner product, a coupling of vectors needs to satisfy:

1) $\langle \underline{a}, \underline{b} \rangle = \overline{\langle \underline{b}, \underline{a} \rangle}$ where the line over the top denotes the conjugate.

2) $\langle \lambda\underline{a}, \underline{b} \rangle = \lambda\langle \underline{a}, \underline{b} \rangle : \lambda \in \mathbb{R}$

3) $\langle \underline{a} + \underline{c}, \underline{b} \rangle = \langle \underline{a}, \underline{b} \rangle + \langle \underline{c}, \underline{b} \rangle$

4) $\langle \underline{a}, \underline{a} \rangle \geq 0$ & $\langle \underline{a}, \underline{a} \rangle = 0$ only when $\underline{a} = 0$

Appendix 2

The Field Axioms

We take a set of mathematical objects. This set of mathematical objects together with operations of addition and multiplication are an algebraic field if they satisfy the following axioms.

The Field Axioms

F1:	Addition is commutative
F2:	Addition is associative
F3:	Inclusion of the additive identity
F4:	Inclusion of additive inverses
F5:	Additive closure
F6:	Multiplicative closure
F7:	Inclusion of multiplicative inverses
F8:	Inclusion of the multiplicative identity
F9:	Multiplication is associative
F10:	Multiplication is distributive over addition
F11:	Multiplication is commutative
F12:	The multiplicative identity is not the same as the additive identity
F13:	The product of the additive identity with any object in the set is the additive identity
F14:	There are no zero divisors

Inclusion of the additive identity means that the additive identity is one of the mathematical objects in the set; similar considerations apply when-ever the word 'inclusion' appears. Additive closure means that the set of mathematical objects includes the sum of any two of the mathematical objects; similar considerations apply when-ever the word 'closure' appears. Zero divisors are non-zero objects that when multiplied together produce the additive identity, zero; for example:

$$\begin{bmatrix} 4 & -4 \\ 2 & -2 \end{bmatrix}\begin{bmatrix} 1 & 3 \\ 1 & 3 \end{bmatrix} = \begin{bmatrix} 0 & 0 \\ 0 & 0 \end{bmatrix}$$

Within this Work:

In our case, this set of mathematical objects is always a set of matrices, and the operations of addition and multiplication are always matrix addition and matrix multiplication. Thus, in our case, we need not consider many of the above axioms because sets of matrices always satisfy them.

The determinant of the zero matrix is zero, and, since determinants are such that $\det(A)\det(B) = \det(AB)$, we can avoid zero divisors by avoiding matrices with zero determinant – singular matrices. (We also need to avoid singular matrices to include multiplicative inverses.) Thus, within this work, we need consider only:

178

The Axioms of an Algebraic Matrix Field[77]

F3: Inclusion of the additive identity, $\begin{bmatrix} 0 & 0 \\ 0 & 0 \end{bmatrix}$

F4: Inclusion of additive inverses
F5: Additive Closure
F6: Multiplicative Closure
F7: Inclusion of multiplicative inverses

F8: Inclusion of the multiplicative identity, $\begin{bmatrix} 1 & 0 \\ 0 & 1 \end{bmatrix}$.

F11: Multiplication is commutative
F14: There are no non-singular matrices in the set

If F11 is satisfied, then we can always calculate the polar form of an algebra. The nature of the polar form is such that it allows no singular matrices, and so we can define the algebra to be the set of matrices that equate to the set of polar form matrices and ignore F14. The polar form automatically includes the multiplicative identity, multiplicative inverses and is multiplicatively closed; thus, if F11 is satisfied, we can ignore F6, F7, and F8.

---

[77] We demonstrate with 2×2 matrices, but the same applies to larger matrices

Appendix 3

General Flat Algebraic Matrix Fields

This is a list of the general form of all flat algebraic matrix forms for $n \times n$ matrices for n from 2 to 4. We also include some general forms for $n = 5, 6$. This includes only the algebraic matrix forms that have n independent variables. We ignore the restrictions necessary to avoid singular matrices; in polar form, any necessary restrictions are automatically forced on to the algebra.

The names appelled to these algebras are formed by:

i) The name of the group that underlies the algebra. If there are more than one algebra derived from a particular group (because sub-groups are generated by different elements of the group), then these different forms are denoted by a decimal point and a number added to the name of the group. Eg: $C_{4.1}$

ii) The polar form of the algebra (if known). Eg: $L^1 H^3$. The L is the (always present) length matrix. The H stands for a rotation matrix with H-type simple trigonometric functions within it. The E stands for a rotation matrix with E-type simple trigonometric functions within it.

iii) The subscripted values of the space squeezing parameters that generate the flat algebra. Eg: $_{(j=1,k=-1)}$

2 Dimensions:
The general 2-dimensional algebraic matrix form is:

$$\begin{bmatrix} a & b \\ jb & a \end{bmatrix}$$

From which we get two flat algebras by taking $j = \{+1, -1\}$:

$$\mathbb{C} = \begin{bmatrix} a & b \\ -b & a \end{bmatrix}, \mathbb{S} = \begin{bmatrix} a & b \\ b & a \end{bmatrix}$$

Calculation of 3-dimensional General Form:
We start with:

$$\begin{bmatrix} a & b & c \\ jc & a & kb \\ lb & mc & a \end{bmatrix}^2 = \begin{bmatrix} a^2 + bc(j+l) & & \\ & a^2 + bc(j+km) & \\ & & a^2 + bc(l+km) \end{bmatrix}$$

We require the second matrix to be of the same form as the first matrix. This gives: $l = j, m = \dfrac{j}{k}$

---

3 Dimensions:

The general 3-dimensional algebraic matrix form is:

$$C_3 = \begin{bmatrix} a & b & b \\ jc & a & kb \\ jb & \dfrac{j}{k}c & a \end{bmatrix} : k \neq 0$$

From which we get four flat algebras by taking $\{j,k\} = \{+1,-1\}$:

$$C_3 L^1 H^2_{(j=1,k=1)} = \begin{bmatrix} a & b & c \\ c & a & b \\ b & c & a \end{bmatrix}, \qquad C_3 L^1 E^2_{(j=1,k=-1)} = \begin{bmatrix} a & b & c \\ c & a & -b \\ b & -c & a \end{bmatrix}$$

$$C_3 L^1 E^1 H^1_{(j=-1,k=1)} = \begin{bmatrix} a & b & c \\ -c & a & b \\ -b & -c & a \end{bmatrix}, \qquad C_3 L^1 E^1 H^1_{(j=-1,k=-1)} = \begin{bmatrix} a & b & c \\ -c & a & -b \\ -b & c & a \end{bmatrix}$$

4 Dimensions:

The 0.1 general 4-dimensional algebraic matrix form with is:

$$C_{4.1} = \begin{bmatrix} a & b & c & d \\ jd & a & kb & lc \\ \dfrac{jl}{k}c & \dfrac{j}{k}d & a & lb \\ jb & \dfrac{j}{k}c & \dfrac{j}{l}d & a \end{bmatrix} : \{k,l\} \neq 0$$

From which we get eight flat algebras by taking $\{j,k,l\} = \{+1,-1\}$:

$$C_{4.1} L^1 H^3_{(j=1,k=1,l=1)} = \begin{bmatrix} a & b & c & d \\ d & a & b & c \\ c & d & a & b \\ b & c & d & a \end{bmatrix}, \qquad C_{4.1} L^1 E^3_{(j=1,k=1,l=-1)} = \begin{bmatrix} a & b & c & d \\ d & a & b & -c \\ -c & d & a & -b \\ b & c & -d & a \end{bmatrix}$$

$$C_{4.1} L^1 E^3_{(j=1,k=-1,l=1)} = \begin{bmatrix} a & b & c & d \\ d & a & -b & c \\ -c & -d & a & b \\ b & -c & d & a \end{bmatrix}, \qquad C_{4.1} L^1 H^3_{(j=1,k=-1,l=-1)} = \begin{bmatrix} a & b & c & d \\ d & a & -b & -c \\ c & -d & a & -b \\ b & -c & -d & a \end{bmatrix}$$

$$C_{4.1}L^1E^3_{(j=-1,k=1,l=1)} = \begin{bmatrix} a & b & c & d \\ -d & a & b & c \\ -c & -d & a & b \\ -b & -c & -d & a \end{bmatrix}, \quad C_{4.1}L^1H^3_{(j=-1,k=1,l=-1)} = \begin{bmatrix} a & b & c & d \\ -d & a & b & -c \\ c & -d & a & -b \\ -b & -c & d & a \end{bmatrix}$$

$$C_{4.1}L^1H^3_{(j=-1,k=-1,l=1)} = \begin{bmatrix} a & b & c & d \\ -d & a & -b & c \\ c & d & a & b \\ -b & c & -d & a \end{bmatrix}, \quad C_{4.1}L^1E^3_{(j=-1,k=-1,l=-1)} = \begin{bmatrix} a & b & c & d \\ -d & a & -b & -c \\ -c & d & a & -b \\ -b & c & d & a \end{bmatrix}$$

The 0.2 general 4-dimensional algebraic matrix form is:

$$C_{4.2} = \begin{bmatrix} a & b & c & d \\ jb & a & kd & \dfrac{j}{k}c \\ ld & \dfrac{l}{k}c & a & \dfrac{j}{k}b \\ lc & \dfrac{kl}{j}d & kb & a \end{bmatrix} : \{j,k\} \neq 0 \in \mathbb{R}$$

From which we get eight flat algebras by taking $\{j,k,l\} = \{+1,-1\}$:

$$C_{4.2}L^1_{(j=1,k=1,l=1)} = \begin{bmatrix} a & b & c & d \\ b & a & d & c \\ d & c & a & b \\ c & d & b & a \end{bmatrix}, \quad C_{4.2}L^1_{(j=-1,k=1,l=1)} = \begin{bmatrix} a & b & c & d \\ -b & a & d & -c \\ d & c & a & -b \\ c & -d & b & a \end{bmatrix}$$

$$C_{4.2}L^1_{(j=1,k=-1,l=1)} = \begin{bmatrix} a & b & c & d \\ b & a & -d & -c \\ d & -c & a & -b \\ c & -d & -b & a \end{bmatrix}, \quad C_{4.2}L^1_{(j=1,k=1,l=-1)} = \begin{bmatrix} a & b & c & d \\ b & a & d & c \\ -d & -c & a & b \\ -c & -d & b & a \end{bmatrix}$$

$$C_{4.2}L^1_{(j=-1,k=-1,l=1)} = \begin{bmatrix} a & b & c & d \\ -b & a & -d & c \\ d & -c & a & b \\ c & d & -b & a \end{bmatrix}, \quad C_{4.2}L^1_{(j=-1,k=1,l=-1)} = \begin{bmatrix} a & b & c & d \\ -b & a & d & -c \\ -d & -c & a & -b \\ -c & d & b & a \end{bmatrix}$$

$$C_{4.2}L^1_{(j=1,k=-1,l=-1)} = \begin{bmatrix} a & b & c & d \\ b & a & -d & -c \\ -d & c & a & -b \\ -c & d & -b & a \end{bmatrix}, \quad C_{4.2}L^1_{(j=-1,k=-1,l=-1)} = \begin{bmatrix} a & b & c & d \\ -b & a & -d & c \\ -d & c & a & b \\ -c & -d & -b & a \end{bmatrix}$$

182

The 0.3 general 4-dimensional algebraic matrix form is:

$$C_{4.3} = \begin{bmatrix} a & b & c & d \\ jc & a & kd & lb \\ jb & \dfrac{j}{l}d & a & \dfrac{j}{k}c \\ \dfrac{jk}{l}d & \dfrac{j}{l}c & kb & a \end{bmatrix} : \{k,l\} \neq 0 \in \mathbb{R}$$

From which we get eight flat algebras by taking $\{j,k,l\} = \{+1,-1\}$:

$$C_{4.3}L^1_{(j=1,k=1,l=1)} = \begin{bmatrix} a & b & c & d \\ c & a & d & b \\ b & d & a & c \\ d & c & b & a \end{bmatrix},
\qquad
C_{4.3}L^1_{(j=-1,k=1,l=1)} = \begin{bmatrix} a & b & c & d \\ -c & a & d & b \\ -b & -d & a & -c \\ -d & -c & -b & a \end{bmatrix}$$

$$C_{4.3}L^1_{(j=1,k=-1,l=1)} = \begin{bmatrix} a & b & c & d \\ c & a & -d & b \\ b & d & a & -c \\ -d & c & -b & a \end{bmatrix},
\qquad
C_{4.3}L^1_{(j=1,k=1,l=-1)} = \begin{bmatrix} a & b & c & d \\ c & a & d & -b \\ b & -d & a & -c \\ -d & -c & b & a \end{bmatrix}$$

$$C_{4.3}L^1_{(j=-1,k=-1,l=1)} = \begin{bmatrix} a & b & c & d \\ -c & a & -d & b \\ -b & -d & a & c \\ d & -c & -b & a \end{bmatrix},
\qquad
C_{4.3}L^1_{(j=-1,k=1,l=-1)} = \begin{bmatrix} a & b & c & d \\ -c & a & d & -b \\ -b & d & a & -c \\ d & c & b & a \end{bmatrix}$$

$$C_{4.3}L^1_{(j=1,k=-1,l=-1)} = \begin{bmatrix} a & b & c & d \\ c & a & -d & -b \\ b & -d & a & -c \\ d & -c & -b & a \end{bmatrix},
\qquad
C_{4.3}L^1_{(j=-1,k=-1,l=-1)} = \begin{bmatrix} a & b & c & d \\ -c & a & -d & -b \\ -b & d & a & c \\ -d & -c & -b & a \end{bmatrix}$$

The general 4-dimensional algebraic matrix form with basic group $C_2 \times C_2$ is:

$$C_2 \times C_2 = \begin{bmatrix} a & b & c & d \\ jb & a & \dfrac{j}{l}d & lc \\ kc & \dfrac{k}{l}d & a & lb \\ \dfrac{jk}{l^2}d & \dfrac{k}{l}c & \dfrac{j}{l}b & a \end{bmatrix} : \{l\} \neq 0$$

From which we get eight flat algebras by taking $\{j,k,l\}=\{+1,-1\}$:

$$C_2 \times C_2 L^1_{(j=1,k=1,l=1)} = \begin{bmatrix} a & b & c & d \\ b & a & d & c \\ c & d & a & b \\ d & c & b & a \end{bmatrix}, \qquad C_2 \times C_2 L^1_{(j=-1,k=1,l=1)} = \begin{bmatrix} a & b & c & d \\ -b & a & -d & c \\ c & d & a & b \\ -d & c & -b & a \end{bmatrix}$$

$$C_2 \times C_2 L^1_{(j=1,k=-1,l=1)} = \begin{bmatrix} a & b & c & d \\ b & a & d & c \\ -c & -d & a & b \\ -d & -c & b & a \end{bmatrix}, \qquad C_2 \times C_2 L^1_{(j=1,k=1,l=-1)} = \begin{bmatrix} a & b & c & d \\ b & a & -d & -c \\ c & -d & a & -b \\ d & -c & -b & a \end{bmatrix}$$

$$C_2 \times C_2 L^1_{(j=-1,k=-1,l=1)} = \begin{bmatrix} a & b & c & d \\ -b & a & -d & c \\ -c & -d & a & b \\ d & -c & -b & a \end{bmatrix}, \qquad C_2 \times C_2 L^1_{(j=-1,k=1,l=-1)} = \begin{bmatrix} a & b & c & d \\ -b & a & d & -c \\ c & -d & a & -b \\ -d & -c & b & a \end{bmatrix}$$

$$C_2 \times C_2 L^1_{(j=1,k=-1,l=-1)} = \begin{bmatrix} a & b & c & d \\ b & a & -d & -c \\ -c & d & a & -b \\ -d & c & -b & a \end{bmatrix}, \qquad C_2 \times C_2 L^1_{(j=-1,k=-1,l=-1)} = \begin{bmatrix} a & b & c & d \\ -b & a & d & -c \\ -c & d & a & -b \\ d & c & b & a \end{bmatrix}$$

5 Dimensions:

The 0.1 general 5-dimensional algebraic matrix form is:

$$C_{5.1} = \begin{bmatrix} a & b & c & d & e \\ je & a & kb & lc & md \\ \dfrac{jm}{k}d & \dfrac{j}{k}e & a & lb & \dfrac{lm}{k}c \\ \dfrac{jm}{k}c & \dfrac{jm}{kl}d & \dfrac{j}{l}e & a & mb \\ jb & \dfrac{j}{k}c & \dfrac{j}{l}d & \dfrac{j}{m}e & a \end{bmatrix} : \{k,l,m \neq 0\}$$

From which we get sixteen flat algebras by taking $\{j,k,l,m\}=\{+1,-1\}$:

6 Dimensions:

The 0.1 general 6-dimensional algebraic matrix form is:

$$C_{6.1} = \begin{bmatrix} a & b & c & d & e & f \\ jf & a & kb & lc & md & ne \\ \dfrac{jn}{k}e & \dfrac{j}{k}f & a & lb & \dfrac{lm}{k}c & \dfrac{mn}{k}d \\ \dfrac{jmn}{kl}d & \dfrac{jn}{kl}e & \dfrac{j}{l}f & a & mb & \dfrac{mn}{k}c \\ \dfrac{jn}{k}c & \dfrac{jn}{kl}d & \dfrac{jn}{lm}e & \dfrac{j}{m}f & a & nb \\ jb & \dfrac{j}{k}c & \dfrac{j}{l}d & \dfrac{j}{m}e & \dfrac{j}{n}f & a \end{bmatrix} : \{k,l,m,n \neq 0\}$$

From which we get thirty-two flat algebras by taking $\{j,k,l,m,n\} = \{+1,-1\}$.

The 0.2 general 6-dimensional algebraic matrix form is:

$$C_{6.2} = \begin{bmatrix} a & b & c & d & e & f \\ jb & a & kd & \dfrac{j}{k}c & lf & \dfrac{j}{l}e \\ me & \dfrac{lm}{j}f & a & \dfrac{j}{k}b & nc & \dfrac{kn}{l}d \\ \dfrac{klm}{j}f & \dfrac{km}{j}e & kb & a & \dfrac{k^2n}{j}d & \dfrac{kn}{l}c \\ mc & \dfrac{km}{j}d & \dfrac{m}{n}e & \dfrac{lm}{kn}f & a & \dfrac{j}{l}b \\ \dfrac{klm}{j}d & \dfrac{lm}{j}c & \dfrac{l^2m}{jn}f & \dfrac{lm}{kn}e & lb & a \end{bmatrix} : \{j,k,l,n \neq 0\}$$

Appendix 4

Circulant Matrices

A $n \times n$ matrix is circulant if it commutes with the appropriately sized forward shift permutation matrix. These are permutation matrices that are effectively the identity matrix shifted one place to the right. For example:

$$\left\{ \begin{bmatrix} 0 & 1 & 0 \\ 0 & 0 & 1 \\ 1 & 0 & 0 \end{bmatrix}, \begin{bmatrix} 0 & 1 & 0 & 0 \\ 0 & 0 & 1 & 0 \\ 0 & 0 & 0 & 1 \\ 1 & 0 & 0 & 0 \end{bmatrix} \right\}$$

of which we have given the 3×3 and the 4×4 versions. The $n \times n$ forward shift matrix is a nth root of unity. Circulant matrices are 'all diagonals'; for example:

$$\left\{ \begin{bmatrix} a & b \\ b & a \end{bmatrix}, \begin{bmatrix} a & b & c \\ c & a & b \\ b & c & a \end{bmatrix}, \begin{bmatrix} a & b & c & d \\ d & a & b & c \\ c & d & a & b \\ b & c & d & a \end{bmatrix} \right\}$$

A $n \times n$ matrix is skew-circulant if it commutes with the skew forward shift matrix. The 3×3 and the 4×4 versions of skew forward shift matrices are:

$$\left\{ \begin{bmatrix} 0 & 1 & 0 \\ 0 & 0 & 1 \\ -1 & 0 & 0 \end{bmatrix}, \begin{bmatrix} 0 & 1 & 0 & 0 \\ 0 & 0 & 1 & 0 \\ 0 & 0 & 0 & 1 \\ -1 & 0 & 0 & 0 \end{bmatrix} \right\}$$

The $n \times n$ skew forward shift matrix is a nth root of minus unity. Examples of skew circulant matrices are:

$$\left\{ \begin{bmatrix} a & b \\ -b & a \end{bmatrix}, \begin{bmatrix} a & b & c \\ -c & a & b \\ -b & -c & a \end{bmatrix}, \begin{bmatrix} a & b & c & d \\ -d & a & b & c \\ -c & -d & a & b \\ -b & -c & -d & a \end{bmatrix} \right\}$$

The standard form Cayley tables of all the cyclic groups have a 'all diagonals' form. When this table is mapped into a matrix, it becomes the circulant algebraic matrix form associated with that order cyclic group. Thus, we can write out an algebraic matrix form of a cyclic group by simply writing out an 'all diagonals' square matrix. Unfortunately, we miss the algebraically isomorphic algebraic matrix forms[78]. We also miss the algebraic matrix forms associated with the non-cyclic abelian groups.

The determinant of the 2×2 circulant matrix is:

---

[78] The $\{C_{42}, C_{43}, C_2 \times C_2\}$ algebraic matrix forms do not commute with the forward shift permutation matrix.

$$\det\left(\begin{bmatrix} a & b \\ b & a \end{bmatrix}\right) = \left(\left(\left(\sqrt[2]{+1}\right)_1\right)^2 a + \left(\sqrt[2]{+1}\right)_1 b\right)\left(\left(\left(\sqrt[2]{+1}\right)_2\right)^2 a + \left(\sqrt[2]{+1}\right)_2 b\right)$$

where $\left(\sqrt[2]{+1}\right)_1 = +1$ is the first square root of unity and $\left(\sqrt[2]{+1}\right)_2 = -1$ is the second square root of unity. The determinant of the 3×3 circulant matrix is:

$$\det\left(\begin{bmatrix} a & b & c \\ c & a & b \\ b & c & a \end{bmatrix}\right) = PROD\begin{pmatrix} \left(\left(\left(\sqrt[3]{+1}\right)_1\right)^3 a + \left(\left(\sqrt[3]{+1}\right)_1\right)^1 b + \left(\left(\sqrt[3]{+1}\right)_1\right)^2 c\right) \\ \left(\left(\left(\sqrt[3]{+1}\right)_2\right)^3 a + \left(\left(\sqrt[3]{+1}\right)_2\right)^1 b + \left(\left(\sqrt[3]{+1}\right)_2\right)^2 c\right) \\ \left(\left(\left(\sqrt[3]{+1}\right)_3\right)^3 a + \left(\left(\sqrt[3]{+1}\right)_3\right)^1 b + \left(\left(\sqrt[3]{+1}\right)_3\right)^2 c\right) \end{pmatrix}$$

where $\left(\sqrt[3]{+1}\right)_1 = +1$ is the first cube root of unity, $\left(\sqrt[3]{+1}\right)_2 = -\dfrac{1}{2}+\dfrac{\sqrt[2]{3}}{2}\hat{i}$ is the second cube

root of unity, and $\left(\sqrt[3]{+1}\right)_3 = -\dfrac{1}{2}-\dfrac{\sqrt[2]{3}}{2}\hat{i}$ is the third cube root of unity. (Of course, these complex numbers do not exist in the 3-dimensional algebras.) The determinant of the 4×4 circulant matrix is:

$$\det\left(\begin{bmatrix} a & b & c & d \\ d & a & b & c \\ c & d & a & b \\ b & c & d & a \end{bmatrix}\right)$$

$$= PROD\begin{pmatrix} \left(\left(\left(\sqrt[4]{+1}\right)_1\right)^4 a + \left(\left(\sqrt[4]{+1}\right)_1\right)^1 b + \left(\left(\sqrt[4]{+1}\right)_1\right)^2 c + \left(\left(\sqrt[4]{+1}\right)_1\right)^3 d\right) \\ \left(\left(\left(\sqrt[4]{+1}\right)_2\right)^4 a + \left(\left(\sqrt[4]{+1}\right)_2\right)^1 b + \left(\left(\sqrt[4]{+1}\right)_2\right)^2 c + \left(\left(\sqrt[4]{+1}\right)_2\right)^3 d\right) \\ \left(\left(\left(\sqrt[4]{+1}\right)_3\right)^4 a + \left(\left(\sqrt[4]{+1}\right)_3\right)^1 b + \left(\left(\sqrt[4]{+1}\right)_3\right)^2 c + \left(\left(\sqrt[4]{+1}\right)_3\right)^3 d\right) \\ \left(\left(\left(\sqrt[4]{+1}\right)_4\right)^4 a + \left(\left(\sqrt[4]{+1}\right)_4\right)^1 b + \left(\left(\sqrt[4]{+1}\right)_4\right)^2 c + \left(\left(\sqrt[4]{+1}\right)_4\right)^3 d\right) \end{pmatrix}$$

$$= PROD\begin{pmatrix} (a+b+c+d) \\ (a-b+c-d) \\ (a+\hat{i}b-c-\hat{i}d) \\ (a-\hat{i}b-c+\hat{i}d) \end{pmatrix}$$

The pattern is clear and it applies to all circulant matrices.

Theorem:[79]
The determinant of the $n \times n$ circulant matrix is:

$$\prod_{j=1}^{j=n} \left(a\left(\omega_j\right)^n + b\left(\omega_j\right)^1 + c\left(\omega_j\right)^2 + d\left(\omega_j\right)^3 + ... + \text{variable}_n \left(\omega_j\right)^{n-1} \right)$$

The eigenvalues of the $n \times n$ circulant matrix are:

$$\lambda_j = a\left(\omega_j\right)^n + b\left(\omega_j\right)^1 + c\left(\omega_j\right)^2 + d\left(\omega_j\right)^3 + ... + \text{variable}_n \left(\omega_j\right)^{n-1}$$

Clearly, one basic form of the algebraic matrix forms of cyclic groups is a circulant matrix.

---

[79] See http://mathworld.wolfram.com/CirculantDeterminant.html

Appendix 5.1

Polar Forms of the $C_{4.1}$ Algebras

The general 4-dimensional algebraic matrix form of the $C_{4.1}$ algebras is:

$$
\begin{bmatrix}
a & b & c & d \\
jd & a & kb & lc \\
\dfrac{jl}{k}c & \dfrac{j}{k}d & a & lb \\
jb & \dfrac{j}{k}c & \dfrac{j}{l}d & a
\end{bmatrix} : \{k,l\} \neq 0
$$

From which we get eight algebras by taking $\{j,k,l\} = \{+1,-1\}$:

$$
\begin{bmatrix}
a & b & c & d \\
d & a & b & c \\
c & d & a & b \\
b & c & d & a
\end{bmatrix},
\begin{bmatrix}
a & b & c & d \\
d & a & b & -c \\
-c & d & a & -b \\
b & c & -d & a
\end{bmatrix},
\begin{bmatrix}
a & b & c & d \\
d & a & -b & c \\
-c & -d & a & b \\
b & -c & d & a
\end{bmatrix},
\begin{bmatrix}
a & b & c & d \\
d & a & -b & -c \\
c & -d & a & -b \\
b & -c & -d & a
\end{bmatrix}
$$

$$
\begin{bmatrix}
a & b & c & d \\
-d & a & b & c \\
-c & -d & a & b \\
-b & -c & -d & a
\end{bmatrix},
\begin{bmatrix}
a & b & c & d \\
-d & a & b & -c \\
c & -d & a & -b \\
-b & -c & d & a
\end{bmatrix},
\begin{bmatrix}
a & b & c & d \\
-d & a & -b & c \\
c & d & a & b \\
-b & c & -d & a
\end{bmatrix},
\begin{bmatrix}
a & b & c & d \\
-d & a & -b & -c \\
-c & d & a & -b \\
-b & c & d & a
\end{bmatrix}
$$

We have:

$$
\exp\left(\begin{bmatrix}
a & b & c & d \\
d & a & b & c \\
c & d & a & b \\
b & c & d & a
\end{bmatrix}\right) = PROD
\left(
\begin{bmatrix}
h & 0 & 0 & 0 \\
0 & h & 0 & 0 \\
0 & 0 & h & 0 \\
0 & 0 & 0 & h
\end{bmatrix}
\begin{bmatrix}
AH_4(b) & BH_4(b) & CH_4(b) & DH_4(b) \\
DH_4(b) & AH_4(b) & BH_4(b) & CH_4(b) \\
CH_4(b) & DH_4(b) & AH_4(b) & BH_4(b) \\
BH_4(b) & CH_4(b) & DH_4(b) & AH_4(b)
\end{bmatrix}
\right.
$$

$$
\begin{bmatrix}
\cosh(c) & 0 & \sinh(c) & 0 \\
0 & \cosh(c) & 0 & \sinh(c) \\
\sinh(c) & 0 & \cosh(c) & 0 \\
0 & \sinh(c) & 0 & \cosh(c)
\end{bmatrix}
$$

$$
\left.
\begin{bmatrix}
AH_4(d) & DH_4(d) & CH_4(d) & BH_4(d) \\
BH_4(d) & AH_4(d) & DH_4(d) & CH_4(d) \\
CH_4(d) & BH_4(d) & AH_4(d) & DH_4(d) \\
DH_4(d) & CH_4(d) & BH_4(d) & AH_4(d)
\end{bmatrix}
\right)
$$

$$\exp\left(\begin{bmatrix} a & b & c & d \\ d & a & b & -c \\ -c & d & a & -b \\ b & c & -d & a \end{bmatrix}\right) = PROD\left(\begin{bmatrix} h & 0 & 0 & 0 \\ 0 & h & 0 & 0 \\ 0 & 0 & h & 0 \\ 0 & 0 & 0 & h \end{bmatrix}\begin{bmatrix} AE_4(b) & BE_4(b) & CE_4(b) & -DE_4(b) \\ -DE_4(b) & AE_4(b) & BE_4(b) & -CE_4(b) \\ -CE_4(b) & -DE_4(b) & AE_4(b) & -BE_4(b) \\ BE_4(b) & CE_4(b) & DE_4(b) & AE_4(b) \end{bmatrix}\right.$$

$$\begin{bmatrix} \cos(c) & 0 & \sin(c) & 0 \\ 0 & \cos(c) & 0 & -\sin(c) \\ -\sin(c) & 0 & \cos(c) & 0 \\ 0 & \sin(c) & 0 & \cos(c) \end{bmatrix}$$

$$\left.\begin{bmatrix} AE_4(d) & -DE_4(d) & -CE_4(d) & BE_4(d) \\ BE_4(d) & AE_4(d) & -DE_4(d) & CE_4(d) \\ CE_4(d) & BE_4(d) & AE_4(d) & DE_4(d) \\ -DE_4(d) & -CE_4(d) & -BE_4(d) & AE_4(d) \end{bmatrix}\right)$$

$$\exp\left(\begin{bmatrix} a & b & c & d \\ d & a & -b & c \\ -c & -d & a & b \\ b & -c & d & a \end{bmatrix}\right) = PROD\left(\begin{bmatrix} h & 0 & 0 & 0 \\ 0 & h & 0 & 0 \\ 0 & 0 & h & 0 \\ 0 & 0 & 0 & h \end{bmatrix}\begin{bmatrix} AE_4(b) & BE_4(b) & -CE_4(b) & -DE_4(b) \\ -DE_4(b) & AE_4(b) & -BE_4(b) & -CE_4(b) \\ CE_4(b) & DE_4(b) & AE_4(b) & BE_4(b) \\ BE_4(b) & CE_4(b) & -DE_4(b) & AE_4(b) \end{bmatrix}\right.$$

$$\begin{bmatrix} \cos(c) & 0 & \sin(c) & 0 \\ 0 & \cos(c) & 0 & \sin(c) \\ -\sin(c) & 0 & \cos(c) & 0 \\ 0 & -\sin(c) & 0 & \cos(c) \end{bmatrix}$$

$$\left.\begin{bmatrix} AE_4(d) & -DE_4(d) & CE_4(d) & BE_4(d) \\ BE_4(d) & AE_4(d) & DE_4(d) & CE_4(d) \\ -CE_4(d) & -BE_4(d) & AE_4(d) & -DE_4(d) \\ -DE_4(d) & -CE_4(d) & BE_4(d) & AE_4(d) \end{bmatrix}\right)$$

$$\exp\left(\begin{bmatrix} a & b & c & d \\ d & a & -b & -c \\ c & -d & a & -b \\ b & -c & -d & a \end{bmatrix}\right) = PROD\left(\begin{bmatrix} h & 0 & 0 & 0 \\ 0 & h & 0 & 0 \\ 0 & 0 & h & 0 \\ 0 & 0 & 0 & h \end{bmatrix}\begin{bmatrix} AH_4(b) & BH_4(b) & -CH_4(b) & DH_4(b) \\ DH_4(b) & AH_4(b) & -BH_4(b) & CH_4(b) \\ -CH_4(b) & -DH_4(b) & AH_4(b) & -BH_4(b) \\ BH_4(b) & CH_4(b) & -DH_4(b) & AH_4(b) \end{bmatrix}\right.$$

$$\begin{bmatrix} \cosh(c) & 0 & \sinh(c) & 0 \\ 0 & \cosh(c) & 0 & -\sinh(c) \\ \sinh(c) & 0 & \cosh(c) & 0 \\ 0 & -\sinh(c) & 0 & \cosh(c) \end{bmatrix}$$

$$\left.\begin{bmatrix} AH_4(d) & DH_4(d) & -CH_4(d) & BH_4(d) \\ BH_4(d) & AH_4(d) & -DH_4(d) & CH_4(d) \\ -CH_4(d) & -BH_4(d) & AH_4(d) & -DH_4(d) \\ DH_4(d) & CH_4(d) & -BH_4(d) & AH_4(d) \end{bmatrix}\right)$$

$$\exp\left(\begin{bmatrix} a & b & c & d \\ -d & a & b & c \\ -c & -d & a & b \\ -b & -c & -d & a \end{bmatrix}\right) = PROD \left(\begin{bmatrix} h & 0 & 0 & 0 \\ 0 & h & 0 & 0 \\ 0 & 0 & h & 0 \\ 0 & 0 & 0 & h \end{bmatrix} \begin{bmatrix} AE_4(b) & BE_4(b) & CE_4(b) & DE_4(b) \\ -DE_4(b) & AE_4(b) & BE_4(b) & CE_4(b) \\ -CE_4(b) & -DE_4(b) & AE_4(b) & BE_4(b) \\ -BE_4(b) & -CE_4(b) & -DE_4(b) & AE_4(b) \end{bmatrix} \right.$$

$$\begin{bmatrix} \cos(c) & 0 & \sin(c) & 0 \\ 0 & \cos(c) & 0 & \sin(c) \\ -\sin(c) & 0 & \cos(c) & 0 \\ 0 & -\sin(c) & 0 & \cos(c) \end{bmatrix}$$

$$\left. \begin{bmatrix} AE_4(d) & DE_4(d) & -CE_4(d) & BE_4(d) \\ -BE_4(d) & AE_4(d) & DE_4(d) & -CE_4(d) \\ CE_4(d) & -BE_4(d) & AE_4(d) & DE_4(d) \\ -DE_4(d) & CE_4(d) & -BE_4(d) & AE_4(d) \end{bmatrix} \right)$$

$$\exp\left(\begin{bmatrix} a & b & c & d \\ -d & a & b & -c \\ c & -d & a & -b \\ -b & -c & d & a \end{bmatrix}\right) = PROD \left(\begin{bmatrix} h & 0 & 0 & 0 \\ 0 & h & 0 & 0 \\ 0 & 0 & h & 0 \\ 0 & 0 & 0 & h \end{bmatrix} \begin{bmatrix} AH_4(b) & BH_4(b) & CH_4(b) & -DH_4(b) \\ DH_4(b) & AH_4(b) & BH_4(b) & -CH_4(b) \\ CH_4(b) & DH_4(b) & AH_4(b) & -BH_4(b) \\ -BH_4(b) & -CH_4(b) & -DH_4(b) & AH_4(b) \end{bmatrix} \right.$$

$$\begin{bmatrix} \cosh(c) & 0 & \sinh(c) & 0 \\ 0 & \cosh(c) & 0 & -\sinh(c) \\ \sinh(c) & 0 & \cosh(c) & 0 \\ 0 & -\sinh(c) & 0 & \cosh(c) \end{bmatrix}$$

$$\left. \begin{bmatrix} AH_4(d) & -DH_4(d) & CH_4(d) & BH_4(d) \\ -BH_4(d) & AH_4(d) & -DH_4(d) & -CH_4(d) \\ CH_4(d) & -BH_4(d) & AH_4(d) & DH_4(d) \\ DH_4(d) & -CH_4(d) & BH_4(d) & AH_4(d) \end{bmatrix} \right)$$

$$\exp\left(\begin{bmatrix} a & b & c & d \\ -d & a & -b & c \\ c & d & a & b \\ -b & c & -d & a \end{bmatrix}\right) = PROD \left(\begin{bmatrix} h & 0 & 0 & 0 \\ 0 & h & 0 & 0 \\ 0 & 0 & h & 0 \\ 0 & 0 & 0 & h \end{bmatrix} \begin{bmatrix} AH_4(b) & BH_4(b) & -CH_4(b) & -DH_4(b) \\ DH_4(b) & AH_4(b) & -BH_4(b) & -CH_4(b) \\ -CH_4(b) & -DH_4(b) & AH_4(b) & BH_4(b) \\ -BH_4(b) & -CH_4(b) & DH_4(b) & AH_4(b) \end{bmatrix} \right.$$

$$\begin{bmatrix} \cosh(c) & 0 & \sinh(c) & 0 \\ \sigma & \cosh(c) & 0 & \sinh(c) \\ \sinh(c) & 0 & \cosh(c) & 0 \\ 0 & \sinh(c) & 0 & \cosh(c) \end{bmatrix}$$

$$\left. \begin{bmatrix} AH_4(d) & -DH_4(d) & -CH_4(d) & BH_4(d) \\ -BH_4(d) & AH_4(d) & DH_4(d) & -CH_4(d) \\ -CH_4(d) & BH_4(d) & AH_4(d) & -DH_4(d) \\ DH_4(d) & -CH_4(d) & -BH_4(d) & AH_4(d) \end{bmatrix} \right)$$

$$\exp\!\left(\begin{bmatrix} a & b & c & d \\ -d & a & -b & -c \\ -c & d & a & -b \\ -b & c & d & a \end{bmatrix}\right) = PROD\left(\begin{bmatrix} h & 0 & 0 & 0 \\ 0 & h & 0 & 0 \\ 0 & 0 & h & 0 \\ 0 & 0 & 0 & h \end{bmatrix}\begin{bmatrix} AE_x(b) & BE_x(b) & -CE_x(b) & DE_x(b) \\ -DE_x(b) & AE_x(b) & -BE_x(b) & CE_x(b) \\ CE_x(b) & DE_x(b) & AE_x(b) & -BE_x(b) \\ -BE_x(b) & -CE_x(b) & DE_x(b) & AE_x(b) \end{bmatrix}\right.$$

$$\begin{bmatrix} \cos(c) & 0 & \sin(c) & 0 \\ 0 & \cos(c) & 0 & -\sin(c) \\ -\sin(c) & 0 & \cos(c) & 0 \\ 0 & \sin(c) & 0 & \cos(c) \end{bmatrix}$$

$$\left.\begin{bmatrix} AE_x(d) & DE_x(d) & CE_x(d) & BE_x(d) \\ -BE_x(d) & AE_x(d) & -DE_x(d) & -CE_x(d) \\ -CE_x(d) & BE_x(d) & AE_x(d) & -DE_x(d) \\ -DE_x(d) & CE_x(d) & BE_x(d) & AE_x(d) \end{bmatrix}\right)$$

Appendix 5.2

Polar Forms of the $C_{4.2}$ Algebras

The general 4-dimensional algebraic matrix form of the $C_{4.2}$ algebras is:

$$\begin{bmatrix} a & b & c & d \\ jb & a & kd & \dfrac{j}{k}c \\ ld & \dfrac{l}{k}c & a & \dfrac{j}{k}b \\ lc & \dfrac{kl}{j}d & kb & a \end{bmatrix} : \{j,k\} \neq 0$$

From which we get eight algebras by taking permutations of $\{j,k,l\} = \{+1,-1\}$:

$$\begin{bmatrix} a & b & c & d \\ b & a & d & c \\ d & c & a & b \\ c & d & b & a \end{bmatrix}, \begin{bmatrix} a & b & c & d \\ -b & a & d & -c \\ d & c & a & -b \\ c & -d & b & a \end{bmatrix}, \begin{bmatrix} a & b & c & d \\ b & a & -d & -c \\ d & -c & a & -b \\ c & -d & -b & a \end{bmatrix}, \begin{bmatrix} a & b & c & d \\ b & a & d & c \\ -d & -c & a & b \\ -c & -d & b & a \end{bmatrix}$$

$$\begin{bmatrix} a & b & c & d \\ -b & a & -d & c \\ d & -c & a & b \\ c & d & -b & a \end{bmatrix}, \begin{bmatrix} a & b & c & d \\ -b & a & d & -c \\ -d & -c & a & -b \\ -c & d & b & a \end{bmatrix}, \begin{bmatrix} a & b & c & d \\ b & a & -d & -c \\ -d & c & a & -b \\ -c & d & -b & a \end{bmatrix}, \begin{bmatrix} a & b & c & d \\ -b & a & -d & c \\ -d & c & a & b \\ -c & -d & -b & a \end{bmatrix}$$

$$\exp\left(\begin{bmatrix} a & b & c & d \\ b & a & d & c \\ d & c & a & b \\ c & d & b & a \end{bmatrix}\right) = PROD\left(\begin{bmatrix} h & 0 & 0 & 0 \\ 0 & h & 0 & 0 \\ 0 & 0 & h & 0 \\ 0 & 0 & 0 & h \end{bmatrix}\begin{bmatrix} \cosh(b) & \sinh(b) & 0 & 0 \\ \sinh(b) & \cosh(b) & 0 & 0 \\ 0 & 0 & \cosh(b) & \sinh(b) \\ 0 & 0 & \sinh(b) & \cosh(b) \end{bmatrix}\right.$$
$$\begin{bmatrix} AH_4(c) & CH_4(c) & BH_4(c) & DH_4(c) \\ CH_4(c) & AH_4(c) & DH_4(c) & BH_4(c) \\ DH_4(c) & BH_4(c) & AH_4(c) & CH_4(c) \\ BH_4(c) & DH_4(c) & CH_4(c) & AH_4(c) \end{bmatrix}$$
$$\left.\begin{bmatrix} AH_4(d) & CH_4(d) & DH_4(d) & BH_4(d) \\ CH_4(d) & AH_4(d) & BH_4(d) & DH_4(d) \\ BH_4(d) & DH_4(d) & AH_4(d) & CH_4(d) \\ DH_4(d) & BH_4(d) & CH_4(d) & AH_4(d) \end{bmatrix}\right)$$

$$\exp\left(\begin{bmatrix} a & b & c & d \\ -b & a & d & -c \\ d & c & a & -b \\ c & -d & b & a \end{bmatrix}\right) = PROD\left(\begin{bmatrix} h & 0 & 0 & 0 \\ 0 & h & 0 & 0 \\ 0 & 0 & h & 0 \\ 0 & 0 & 0 & h \end{bmatrix}\begin{bmatrix} \cos(b) & \sin(b) & 0 & 0 \\ -\sin(b) & \cos(b) & 0 & 0 \\ 0 & 0 & \cos(b) & -\sin(b) \\ 0 & 0 & \sin(b) & \cos(b) \end{bmatrix}\right.$$
$$\begin{bmatrix} AE_4(c) & CE_4(c) & BE_4(c) & -DE_4(c) \\ -CE_4(c) & AE_4(c) & -DE_4(c) & -BE_4(c) \\ -DE_4(c) & BE_4(c) & AE_4(c) & -CE_4(c) \\ BE_4(c) & DE_4(c) & CE_4(c) & AE_4(c) \end{bmatrix}$$
$$\left.\begin{bmatrix} AE_4(d) & -CE_4(d) & -DE_4(d) & BE_4(d) \\ CE_4(d) & AE_4(d) & BE_4(d) & DE_4(d) \\ BE_4(d) & -DE_4(d) & AE_4(d) & CE_4(d) \\ -DE_4(d) & -BE_4(d) & -CE_4(d) & AE_4(d) \end{bmatrix}\right)$$

$$\exp\left(\begin{bmatrix} a & b & c & d \\ b & a & -d & -c \\ d & -c & a & -b \\ c & -d & -b & a \end{bmatrix}\right) = PROD\left(\begin{bmatrix} h & 0 & 0 & 0 \\ 0 & h & 0 & 0 \\ 0 & 0 & h & 0 \\ 0 & 0 & 0 & h \end{bmatrix}\begin{bmatrix} \cosh(b) & \sinh(b) & 0 & 0 \\ \sinh(b) & \cosh(b) & 0 & 0 \\ 0 & 0 & \cosh(b) & -\sinh(b) \\ 0 & 0 & -\sinh(b) & \cosh(b) \end{bmatrix}\right.$$
$$\begin{bmatrix} AH_4(c) & -CH_4(c) & BH_4(c) & DH_4(c) \\ -CH_4(c) & AH_4(c) & -DH_4(c) & -BH_4(c) \\ DH_4(c) & -BH_4(c) & AH_4(c) & CH_4(c) \\ BH_4(c) & -DH_4(c) & CH_4(c) & AH_4(c) \end{bmatrix}$$
$$\left.\begin{bmatrix} AH_4(d) & -CH_4(d) & DH_4(d) & BH_4(d) \\ -CH_4(d) & AH_4(d) & -BH_4(d) & -DH_4(d) \\ BH_4(d) & -DH_4(d) & AH_4(d) & CH_4(d) \\ DH_4(d) & -BH_4(d) & CH_4(d) & AH_4(d) \end{bmatrix}\right)$$

$$\exp\left(\begin{bmatrix} a & b & c & d \\ b & a & d & c \\ -d & -c & a & b \\ -c & -d & b & a \end{bmatrix}\right) = PROD \left(\begin{bmatrix} h & 0 & 0 & 0 \\ 0 & h & 0 & 0 \\ 0 & 0 & h & 0 \\ 0 & 0 & 0 & h \end{bmatrix} \begin{bmatrix} \cosh(b) & \sinh(b) & 0 & 0 \\ \sinh(b) & \cosh(b) & 0 & 0 \\ 0 & 0 & \cosh(b) & \sinh(b) \\ 0 & 0 & \sinh(b) & \cosh(b) \end{bmatrix} \right.$$

$$\begin{bmatrix} AH_4(c) & -CH_4(c) & BH_4(c) & -DH_4(c) \\ -CH_4(c) & AH_4(c) & -DH_4(c) & BH_4(c) \\ DH_4(c) & -BH_4(c) & AH_4(c) & -CH_4(c) \\ -BH_4(c) & DH_4(c) & -CH_4(c) & AH_4(c) \end{bmatrix}$$

$$\left. \begin{bmatrix} AH_4(d) & -CH_4(d) & -DH_4(d) & BH_4(d) \\ -CH_4(d) & AH_4(d) & BH_4(d) & -DH_4(d) \\ -BH_4(d) & DH_4(d) & AH_4(d) & -CH_4(d) \\ DH_4(d) & -BH_4(d) & -CH_4(d) & AH_4(d) \end{bmatrix} \right)$$

$$\exp\left(\begin{bmatrix} a & b & c & d \\ -b & a & -d & c \\ d & -c & a & b \\ c & d & -b & a \end{bmatrix}\right) = PROD \left(\begin{bmatrix} h & 0 & 0 & 0 \\ 0 & h & 0 & 0 \\ 0 & 0 & h & 0 \\ 0 & 0 & 0 & h \end{bmatrix} \begin{bmatrix} \cos(b) & \sin(b) & 0 & 0 \\ -\sin(b) & \cos(b) & 0 & 0 \\ 0 & 0 & \cos(b) & \sin(b) \\ 0 & 0 & -\sin(b) & \cos(b) \end{bmatrix} \right.$$

$$\begin{bmatrix} AE_4(c) & -CE_4(c) & BE_4(c) & -DE_4(c) \\ CE_4(c) & AE_4(c) & DE_4(c) & BE_4(c) \\ -DE_4(c) & -BE_4(c) & AE_4(c) & -CE_4(c) \\ BE_4(c) & -DE_4(c) & CE_4(c) & AE_4(c) \end{bmatrix}$$

$$\left. \begin{bmatrix} AE_4(d) & CE_4(d) & -DE_4(d) & BE_4(d) \\ -CE_4(d) & AE_4(d) & -BE_4(d) & -DE_4(d) \\ BE_4(d) & DE_4(d) & AE_4(d) & CE_4(d) \\ -DE_4(d) & BE_4(d) & -CE_4(d) & AE_4(d) \end{bmatrix} \right)$$

$$\exp\left(\begin{bmatrix} a & b & c & d \\ -b & a & d & -c \\ -d & -c & a & -b \\ -c & d & b & a \end{bmatrix}\right) = PROD \left(\begin{bmatrix} h & 0 & 0 & 0 \\ 0 & h & 0 & 0 \\ 0 & 0 & h & 0 \\ 0 & 0 & 0 & h \end{bmatrix} \begin{bmatrix} \cos(b) & \sin(b) & 0 & 0 \\ -\sin(b) & \cos(b) & 0 & 0 \\ 0 & 0 & \cos(b) & -\sin(b) \\ 0 & 0 & \sin(b) & \cos(b) \end{bmatrix} \right.$$

$$\begin{bmatrix} AE_4(c) & -CE_4(c) & BE_4(c) & DE_4(c) \\ CE_4(c) & AE_4(c) & DE_4(c) & -BE_4(c) \\ -DE_4(c) & -BE_4(c) & AE_4(c) & CE_4(c) \\ -BE_4(c) & DE_4(c) & -CE_4(c) & AE_4(c) \end{bmatrix}$$

$$\left. \begin{bmatrix} AE_4(d) & CE_4(d) & DE_4(d) & BE_4(d) \\ -CE_4(d) & AE_4(d) & BE_4(d) & -DE_4(d) \\ -BE_4(d) & -DE_4(d) & AE_4(d) & -CE_4(d) \\ -DE_4(d) & BE_4(d) & CE_4(d) & AE_4(d) \end{bmatrix} \right)$$

$$\exp\left(\begin{bmatrix} a & b & c & d \\ b & a & -d & -c \\ -d & c & a & -b \\ -c & d & -b & a \end{bmatrix}\right) = PROD \left(\begin{bmatrix} h & 0 & 0 & 0 \\ 0 & h & 0 & 0 \\ 0 & 0 & h & 0 \\ 0 & 0 & 0 & h \end{bmatrix} \begin{bmatrix} \cosh(b) & \sinh(b) & 0 & 0 \\ \sinh(b) & \cosh(b) & 0 & 0 \\ 0 & 0 & \cosh(b) & -\sinh(b) \\ 0 & 0 & -\sinh(b) & \cosh(b) \end{bmatrix} \right.$$

$$\begin{bmatrix} AH_4(c) & CH_4(c) & BH_4(c) & -DH_4(c) \\ CH_4(c) & AH_4(c) & DH_4(c) & -BH_4(c) \\ DH_4(c) & BH_4(c) & AH_4(c) & -CH_4(c) \\ -BH_4(c) & -DH_4(c) & -CH_4(c) & AH_4(c) \end{bmatrix}$$

$$\left.\begin{bmatrix} AH_4(d) & CH_4(d) & -DH_4(d) & BH_4(d) \\ CH_4(d) & AH_4(d) & -BH_4(d) & DH_4(d) \\ -BH_4(d) & -DH_4(d) & AH_4(d) & -CH_4(d) \\ DH_4(d) & BH_4(d) & -CH_4(d) & AH_4(d) \end{bmatrix}\right)$$

$$\exp\left(\begin{bmatrix} a & b & c & d \\ -b & a & -d & c \\ -d & c & a & b \\ -c & -d & -b & a \end{bmatrix}\right) = PROD \left(\begin{bmatrix} h & 0 & 0 & 0 \\ 0 & h & 0 & 0 \\ 0 & 0 & h & 0 \\ 0 & 0 & 0 & h \end{bmatrix} \begin{bmatrix} \cos(b) & \sin(b) & 0 & 0 \\ -\sin(b) & \cos(b) & 0 & 0 \\ 0 & 0 & \cos(b) & \sin(b) \\ 0 & 0 & -\sin(b) & \cos(b) \end{bmatrix} \right.$$

$$\begin{bmatrix} AE_4(c) & CE_4(c) & BE_4(c) & DE_4(c) \\ -CE_4(c) & AE_4(c) & -DE_4(c) & BE_4(c) \\ -DE_4(c) & BE_4(c) & AE_4(c) & CE_4(c) \\ -BE_4(c) & -DE_4(c) & -CE_4(c) & AE_4(c) \end{bmatrix}$$

$$\left.\begin{bmatrix} AE_4(d) & -CE_4(d) & DE_4(d) & BE_4(d) \\ CE_4(d) & AE_4(d) & -BE_4(d) & DE_4(d) \\ -BE_4(d) & DE_4(d) & AE_4(d) & -CE_4(d) \\ -DE_4(d) & -BE_4(d) & CE_4(d) & AE_4(d) \end{bmatrix}\right)$$

Appendix 5.3

Polar Forms of the $C_{4.3}$ Algebras

The general 4-dimensional algebraic matrix form of the $C_{4.3}$ algebras is:

$$\begin{bmatrix} a & b & c & d \\ jc & a & kd & lb \\ jb & \dfrac{j}{l}d & a & \dfrac{j}{k}c \\ \dfrac{jk}{l}d & \dfrac{j}{l}c & kb & a \end{bmatrix} : \{k,l\} \neq 0$$

From which we get eight algebras by taking permutations of $\{j,k,l\} = \{+1,-1\}$:

196

$$\begin{bmatrix} a & b & c & d \\ c & a & d & b \\ b & d & a & c \\ d & c & b & a \end{bmatrix}, \begin{bmatrix} a & b & c & d \\ -c & a & d & b \\ -b & -d & a & -c \\ -d & -c & b & a \end{bmatrix}, \begin{bmatrix} a & b & c & d \\ c & a & -d & b \\ b & d & a & -c \\ -d & c & -b & a \end{bmatrix}, \begin{bmatrix} a & b & c & d \\ c & a & d & -b \\ b & -d & a & c \\ -d & -c & b & a \end{bmatrix}$$

$$\begin{bmatrix} a & b & c & d \\ -c & a & -d & b \\ -b & -d & a & c \\ d & -c & -b & a \end{bmatrix}, \begin{bmatrix} a & b & c & d \\ -c & a & d & -b \\ -b & d & a & -c \\ d & c & b & a \end{bmatrix}, \begin{bmatrix} a & b & c & d \\ c & a & -d & -b \\ b & -d & a & -c \\ d & -c & -b & a \end{bmatrix}, \begin{bmatrix} a & b & c & d \\ -c & a & -d & -b \\ -b & d & a & c \\ -d & c & -b & a \end{bmatrix}$$

$$\exp\left(\begin{bmatrix} a & b & c & d \\ c & a & d & b \\ b & d & a & c \\ d & c & b & a \end{bmatrix}\right) = PROD \left(\begin{bmatrix} h & 0 & 0 & 0 \\ 0 & h & 0 & 0 \\ 0 & 0 & h & 0 \\ 0 & 0 & 0 & h \end{bmatrix} \begin{bmatrix} AH_4(b) & BH_4(b) & DH_4(b) & CH_4(b) \\ DH_4(b) & AH_4(b) & CH_4(b) & BH_4(b) \\ BH_4(b) & CH_4(b) & AH_4(b) & DH_4(b) \\ CH_4(b) & DH_4(b) & BH_4(b) & AH_4(b) \end{bmatrix} \right.$$

$$\begin{bmatrix} AH_4(c) & DH_4(c) & BH_4(c) & CH_4(c) \\ BH_4(c) & AH_4(c) & CH_4(c) & DH_4(c) \\ DH_4(c) & CH_4(c) & AH_4(c) & BH_4(c) \\ CH_4(c) & BH_4(c) & DH_4(c) & AH_4(c) \end{bmatrix}$$

$$\left. \begin{bmatrix} \cosh(d) & 0 & 0 & \sinh(d) \\ 0 & \cosh(d) & \sinh(d) & 0 \\ 0 & \sinh(d) & \cosh(d) & 0 \\ \sinh(d) & 0 & 0 & \cosh(d) \end{bmatrix} \right)$$

$$\exp\left(\begin{bmatrix} a & b & c & d \\ -c & a & d & b \\ -b & -d & a & -c \\ -d & -c & b & a \end{bmatrix}\right) = PROD \left(\begin{bmatrix} h & 0 & 0 & 0 \\ 0 & h & 0 & 0 \\ 0 & 0 & h & 0 \\ 0 & 0 & 0 & h \end{bmatrix} \begin{bmatrix} AE_4(b) & BE_4(b) & DE_4(b) & CE_4(b) \\ -DE_4(b) & AE_4(b) & CE_4(b) & BE_4(b) \\ -BE_4(b) & -CE_4(b) & AE_4(b) & -DE_4(b) \\ -CE_4(b) & -DE_4(b) & BE_4(b) & AE_4(b) \end{bmatrix} \right.$$

$$\begin{bmatrix} AE_4(c) & DE_4(c) & BE_4(c) & -CE_4(c) \\ -BE_4(c) & AE_4(c) & -CE_4(c) & DE_4(c) \\ -DE_4(c) & CE_4(c) & AE_4(c) & -BE_4(c) \\ CE_4(c) & -BE_4(c) & DE_4(c) & AE_4(c) \end{bmatrix}$$

$$\left. \begin{bmatrix} \cos(d) & 0 & 0 & \sin(d) \\ 0 & \cos(d) & \sin(d) & 0 \\ 0 & -\sin(d) & \cos(d) & 0 \\ -\sin(d) & 0 & 0 & \cos(d) \end{bmatrix} \right)$$

$$\exp\left(\begin{bmatrix} a & b & c & d \\ c & a & -d & b \\ b & d & a & -c \\ -d & c & -b & a \end{bmatrix}\right) = PROD\left(\begin{matrix} \begin{bmatrix} h & 0 & 0 & 0 \\ 0 & h & 0 & 0 \\ 0 & 0 & h & 0 \\ 0 & 0 & 0 & h \end{bmatrix}\begin{bmatrix} AE_4(b) & BE_4(b) & -DE_4(b) & CE_4(b) \\ -DE_4(b) & AE_4(b) & -CE_4(b) & BE_4(b) \\ BE_4(b) & CE_4(b) & AE_4(b) & DE_4(b) \\ -CE_4(b) & -DE_4(b) & -BE_4(b) & AE_4(b) \end{bmatrix} \\[2em] \begin{bmatrix} AE_4(c) & -DE_4(c) & BE_4(c) & -CE_4(c) \\ BE_4(c) & AE_4(c) & CE_4(c) & -DE_4(c) \\ -DE_4(c) & -CE_4(c) & AE_4(c) & -BE_4(c) \\ CE_4(c) & BE_4(c) & DE_4(c) & AE_4(c) \end{bmatrix} \\[2em] \begin{bmatrix} \cos(d) & 0 & 0 & \sin(d) \\ 0 & \cos(d) & -\sin(d) & 0 \\ 0 & \sin(d) & \cos(d) & 0 \\ -\sin(d) & 0 & 0 & \cos(d) \end{bmatrix} \end{matrix}\right)$$

$$\exp\left(\begin{bmatrix} a & b & c & d \\ c & a & d & -b \\ b & -d & a & c \\ -d & -c & b & a \end{bmatrix}\right) = PROD\left(\begin{matrix} \begin{bmatrix} h & 0 & 0 & 0 \\ 0 & h & 0 & 0 \\ 0 & 0 & h & 0 \\ 0 & 0 & 0 & h \end{bmatrix}\begin{bmatrix} AE_4(b) & BE_4(b) & -DE_4(b) & -CE_4(b) \\ -DE_4(b) & AE_4(b) & -CE_4(b) & -BE_4(b) \\ BE_4(b) & CE_4(b) & AE_4(b) & -DE_4(b) \\ CE_4(b) & DE_4(b) & BE_4(b) & AE_4(b) \end{bmatrix} \\[2em] \begin{bmatrix} AE_4(c) & -DE_4(c) & BE_4(c) & CE_4(c) \\ BE_4(c) & AE_4(c) & CE_4(c) & DE_4(c) \\ -DE_4(c) & -CE_4(c) & AE_4(c) & BE_4(c) \\ -CE_4(c) & -BE_4(c) & -DE_4(c) & AE_4(c) \end{bmatrix} \\[2em] \begin{bmatrix} \cos(d) & 0 & 0 & \sin(d) \\ 0 & \cos(d) & \sin(d) & 0 \\ 0 & -\sin(d) & \cos(d) & 0 \\ -\sin(d) & 0 & 0 & \cos(d) \end{bmatrix} \end{matrix}\right)$$

$$\exp\left(\begin{bmatrix} a & b & c & d \\ -c & a & -d & b \\ -b & -d & a & c \\ d & -c & -b & a \end{bmatrix}\right) = PROD\left(\begin{matrix} \begin{bmatrix} h & 0 & 0 & 0 \\ 0 & h & 0 & 0 \\ 0 & 0 & h & 0 \\ 0 & 0 & 0 & h \end{bmatrix}\begin{bmatrix} AH_4(b) & BH_4(b) & -DH_4(b) & CH_4(b) \\ DH_4(b) & AH_4(b) & -CH_4(b) & BH_4(b) \\ -BH_4(b) & -CH_4(b) & AH_4(b) & -DH_4(b) \\ CH_4(b) & DH_4(b) & -BH_4(b) & AH_4(b) \end{bmatrix} \\[2em] \begin{bmatrix} AH_4(c) & -DH_4(c) & BH_4(c) & CH_4(c) \\ -BH_4(c) & AH_4(c) & -CH_4(c) & -DH_4(c) \\ DH_4(c) & -CH_4(c) & AH_4(c) & BH_4(c) \\ CH_4(c) & -BH_4(c) & DH_4(c) & AH_4(c) \end{bmatrix} \\[2em] \begin{bmatrix} \cosh(d) & 0 & 0 & \sinh(d) \\ 0 & \cosh(d) & -\sinh(d) & 0 \\ 0 & -\sinh(d) & \cosh(d) & 0 \\ \sinh(d) & 0 & 0 & \cosh(d) \end{bmatrix} \end{matrix}\right)$$

$$\exp\left(\begin{bmatrix} a & b & c & d \\ -c & a & d & -b \\ -b & d & a & -c \\ d & c & b & a \end{bmatrix}\right) = PROD\left(\begin{bmatrix} h & 0 & 0 & 0 \\ 0 & h & 0 & 0 \\ 0 & 0 & h & 0 \\ 0 & 0 & 0 & h \end{bmatrix}\begin{bmatrix} AH_4(b) & BH_4(b) & -DH_4(b) & -CH_4(b) \\ DH_4(b) & AH_4(b) & -CH_4(b) & -BH_4(b) \\ -BH_4(b) & -CH_4(b) & AH_4(b) & DH_4(b) \\ -CH_4(b) & -DH_4(b) & BH_4(b) & AH_4(b) \end{bmatrix}\right.$$

$$\begin{bmatrix} AH_4(c) & -DH_4(c) & BH_4(c) & -CH_4(c) \\ -BH_4(c) & AH_4(c) & -CH_4(c) & DH_4(c) \\ DH_4(c) & -CH_4(c) & AH_4(c) & -BH_4(c) \\ -CH_4(c) & BH_4(c) & -DH_4(c) & AH_4(c) \end{bmatrix}$$

$$\left.\begin{bmatrix} \cosh(d) & 0 & 0 & \sinh(d) \\ 0 & \cosh(d) & \sinh(d) & 0 \\ 0 & \sinh(d) & \cosh(d) & 0 \\ \sinh(d) & 0 & 0 & \cosh(d) \end{bmatrix}\right)$$

$$\exp\left(\begin{bmatrix} a & b & c & d \\ c & a & -d & -b \\ b & -d & a & -c \\ d & -c & -b & a \end{bmatrix}\right) = PROD\left(\begin{bmatrix} h & 0 & 0 & 0 \\ 0 & h & 0 & 0 \\ 0 & 0 & h & 0 \\ 0 & 0 & 0 & h \end{bmatrix}\begin{bmatrix} AH_4(b) & BH_4(b) & DH_4(b) & -CH_4(b) \\ DH_4(b) & AH_4(b) & CH_4(b) & -BH_4(b) \\ BH_4(b) & CH_4(b) & AH_4(b) & -DH_4(b) \\ -CH_4(b) & -DH_4(b) & -BH_4(b) & AH_4(b) \end{bmatrix}\right.$$

$$\begin{bmatrix} AH_4(c) & DH_4(c) & BH_4(c) & -CH_4(c) \\ BH_4(c) & AH_4(c) & CH_4(c) & -DH_4(c) \\ DH_4(c) & CH_4(c) & AH_4(c) & -BH_4(c) \\ -CH_4(c) & -BH_4(c) & -DH_4(c) & AH_4(c) \end{bmatrix}$$

$$\left.\begin{bmatrix} \cosh(d) & 0 & 0 & \sinh(d) \\ 0 & \cosh(d) & -\sinh(d) & 0 \\ 0 & -\sinh(d) & \cosh(d) & 0 \\ \sinh(d) & 0 & 0 & \cosh(d) \end{bmatrix}\right)$$

$$\exp\left(\begin{bmatrix} a & b & c & d \\ -c & a & -d & -b \\ -b & d & a & c \\ -d & c & -b & a \end{bmatrix}\right) = PROD\left(\begin{bmatrix} h & 0 & 0 & 0 \\ 0 & h & 0 & 0 \\ 0 & 0 & h & 0 \\ 0 & 0 & 0 & h \end{bmatrix}\begin{bmatrix} AE_4(b) & BE_4(b) & DE_4(b) & -CE_4(b) \\ -DE_4(b) & AE_4(b) & CE_4(b) & -BE_4(b) \\ -BE_4(b) & -CE_4(b) & AE_4(b) & DE_4(b) \\ CE_4(b) & DE_4(b) & -BE_4(b) & AE_4(b) \end{bmatrix}\right.$$

$$\begin{bmatrix} AE_4(c) & DE_4(c) & BE_4(c) & CE_4(c) \\ -BE_4(c) & AE_4(c) & -CE_4(c) & -DE_4(c) \\ -DE_4(c) & CE_4(c) & AE_4(c) & BE_4(c) \\ -CE_4(c) & BE_4(c) & -DE_4(c) & AE_4(c) \end{bmatrix}$$

$$\left.\begin{bmatrix} \cos(d) & 0 & 0 & \sin(d) \\ 0 & \cos(d) & -\sin(d) & 0 \\ 0 & \sin(d) & \cos(d) & 0 \\ -\sin(d) & 0 & 0 & \cos(d) \end{bmatrix}\right)$$

Appendix 5.4

Polar Forms of the $C_2 \times C_2$ Algebras

The general 4-dimensional algebraic matrix form of the $C_2 \times C_2$ algebras is:

$$\begin{bmatrix} a & b & c & d \\ jb & a & \dfrac{j}{l}d & lc \\ kc & \dfrac{k}{l}d & a & lb \\ \dfrac{jk}{l^2}d & \dfrac{k}{l}c & \dfrac{j}{l}b & a \end{bmatrix} : \{l\} \neq 0$$

From which we get eight flat algebras by taking permutations of $\{j,k,l\} = \{+1,-1\}$:

$$\begin{bmatrix} a & b & c & d \\ b & a & d & c \\ c & d & a & b \\ d & c & b & a \end{bmatrix}, \begin{bmatrix} a & b & c & d \\ -b & a & -d & c \\ c & d & a & b \\ -d & c & -b & a \end{bmatrix}, \begin{bmatrix} a & b & c & d \\ b & a & d & c \\ -c & -d & a & b \\ -d & -c & b & a \end{bmatrix}, \begin{bmatrix} a & b & c & d \\ b & a & -d & -c \\ c & -d & a & -b \\ d & -c & -b & a \end{bmatrix}$$

$$\begin{bmatrix} a & b & c & d \\ -b & a & -d & c \\ -c & -d & a & b \\ d & -c & -b & a \end{bmatrix}, \begin{bmatrix} a & b & c & d \\ -b & a & d & -c \\ c & -d & a & -b \\ -d & -c & b & a \end{bmatrix}, \begin{bmatrix} a & b & c & d \\ b & a & -d & -c \\ -c & d & a & -b \\ -d & c & -b & a \end{bmatrix}, \begin{bmatrix} a & b & c & d \\ -b & a & d & -c \\ -c & d & a & -b \\ d & c & b & a \end{bmatrix}$$

$$\exp\left(\begin{bmatrix} a & b & c & d \\ b & a & d & c \\ c & d & a & b \\ d & c & b & a \end{bmatrix}\right) = PROD \left(\begin{bmatrix} h & 0 & 0 & 0 \\ 0 & h & 0 & 0 \\ 0 & 0 & h & 0 \\ 0 & 0 & 0 & h \end{bmatrix} \begin{bmatrix} \cosh(b) & \sinh(b) & 0 & 0 \\ \sinh(b) & \cosh(b) & 0 & 0 \\ 0 & 0 & \cosh(b) & \sinh(b) \\ 0 & 0 & \sinh(b) & \cosh(b) \end{bmatrix} \right.$$

$$\begin{bmatrix} \cosh(c) & 0 & \sinh(c) & 0 \\ 0 & \cosh(c) & 0 & \sinh(c) \\ \sinh(c) & 0 & \cosh(c) & 0 \\ 0 & \sinh(c) & 0 & \cosh(c) \end{bmatrix}$$

$$\left. \begin{bmatrix} \cosh(d) & 0 & 0 & \sinh(d) \\ 0 & \cosh(d) & \sinh(d) & 0 \\ 0 & \sinh(d) & \cosh(d) & 0 \\ \sinh(d) & 0 & 0 & \cosh(d) \end{bmatrix} \right)$$

$$\exp\left(\begin{bmatrix} a & b & c & d \\ -b & a & -d & c \\ c & d & a & b \\ -d & c & -b & a \end{bmatrix}\right) = PROD \left(\begin{bmatrix} h & 0 & 0 & 0 \\ 0 & h & 0 & 0 \\ 0 & 0 & h & 0 \\ 0 & 0 & 0 & h \end{bmatrix}\begin{bmatrix} \cos(b) & \sin(b) & 0 & 0 \\ -\sin(b) & \cos(b) & 0 & 0 \\ 0 & 0 & \cos(b) & \sin(b) \\ 0 & 0 & -\sin(b) & \cos(b) \end{bmatrix}\right.$$

$$\begin{bmatrix} \cosh(c) & 0 & \sinh(c) & 0 \\ 0 & \cosh(c) & 0 & \sinh(c) \\ \sinh(c) & 0 & \cosh(c) & 0 \\ 0 & \sinh(c) & 0 & \cosh(c) \end{bmatrix}$$

$$\left.\begin{bmatrix} \cos(d) & 0 & 0 & \sin(d) \\ 0 & \cos(d) & -\sin(d) & 0 \\ 0 & \sin(d) & \cos(d) & 0 \\ -\sin(d) & 0 & 0 & \cos(d) \end{bmatrix}\right)$$

$$\exp\left(\begin{bmatrix} a & b & c & d \\ b & a & d & c \\ -c & -d & a & b \\ -d & -c & b & a \end{bmatrix}\right) = PROD \left(\begin{bmatrix} h & 0 & 0 & 0 \\ 0 & h & 0 & 0 \\ 0 & 0 & h & 0 \\ 0 & 0 & 0 & h \end{bmatrix}\begin{bmatrix} \cosh(b) & \sinh(b) & 0 & 0 \\ \sinh(b) & \cosh(b) & 0 & 0 \\ 0 & 0 & \cosh(b) & \sinh(b) \\ 0 & 0 & \sinh(b) & \cosh(b) \end{bmatrix}\right.$$

$$\begin{bmatrix} \cos(c) & 0 & \sin(c) & 0 \\ 0 & \cos(c) & 0 & \sin(c) \\ -\sin(c) & 0 & \cos(c) & 0 \\ 0 & -\sin(c) & 0 & \cos(c) \end{bmatrix}$$

$$\left.\begin{bmatrix} \cos(d) & 0 & 0 & \sin(d) \\ 0 & \cos(d) & \sin(d) & 0 \\ 0 & -\sin(d) & \cos(d) & 0 \\ -\sin(d) & 0 & 0 & \cos(d) \end{bmatrix}\right)$$

$$\exp\left(\begin{bmatrix} a & b & c & d \\ b & a & -d & -c \\ c & -d & a & -b \\ d & -c & -b & a \end{bmatrix}\right) = PROD \left(\begin{bmatrix} h & 0 & 0 & 0 \\ 0 & h & 0 & 0 \\ 0 & 0 & h & 0 \\ 0 & 0 & 0 & h \end{bmatrix}\begin{bmatrix} \cosh(b) & \sinh(b) & 0 & 0 \\ \sinh(b) & \cosh(b) & 0 & 0 \\ 0 & 0 & \cosh(b) & -\sinh(b) \\ 0 & 0 & -\sinh(b) & \cosh(b) \end{bmatrix}\right.$$

$$\begin{bmatrix} \cosh(c) & 0 & \sinh(c) & 0 \\ 0 & \cosh(c) & 0 & -\sinh(c) \\ \sinh(c) & 0 & \cosh(c) & 0 \\ 0 & -\sinh(c) & 0 & \cosh(c) \end{bmatrix}$$

$$\left.\begin{bmatrix} \cosh(d) & 0 & 0 & \sinh(d) \\ 0 & \cosh(d) & -\sinh(d) & 0 \\ 0 & -\sinh(d) & \cosh(d) & 0 \\ \sinh(d) & 0 & 0 & \cosh(d) \end{bmatrix}\right)$$

$$\exp\left(\begin{bmatrix} a & b & c & d \\ -b & a & -d & c \\ -c & -d & a & b \\ d & -c & -b & a \end{bmatrix}\right) = PROD \left(\begin{bmatrix} h & 0 & 0 & 0 \\ 0 & h & 0 & 0 \\ 0 & 0 & h & 0 \\ 0 & 0 & 0 & h \end{bmatrix} \begin{bmatrix} \cos(b) & \sin(b) & 0 & 0 \\ -\sin(b) & \cos(b) & 0 & 0 \\ 0 & 0 & \cos(b) & \sin(b) \\ 0 & 0 & -\sin(b) & \cos(b) \end{bmatrix} \right.$$

$$\begin{bmatrix} \cos(c) & 0 & \sin(c) & 0 \\ 0 & \cos(c) & 0 & \sin(c) \\ -\sin(c) & 0 & \cos(c) & 0 \\ 0 & -\sin(c) & 0 & \cos(c) \end{bmatrix}$$

$$\left.\begin{bmatrix} \cosh(d) & 0 & 0 & \sinh(d) \\ 0 & \cosh(d) & -\sinh(d) & 0 \\ 0 & -\sinh(d) & \cosh(d) & 0 \\ \sinh(d) & 0 & 0 & \cosh(d) \end{bmatrix}\right)$$

$$\exp\left(\begin{bmatrix} a & b & c & d \\ -b & a & d & -c \\ c & -d & a & -b \\ -d & -c & b & a \end{bmatrix}\right) = PROD \left(\begin{bmatrix} h & 0 & 0 & 0 \\ 0 & h & 0 & 0 \\ 0 & 0 & h & 0 \\ 0 & 0 & 0 & h \end{bmatrix} \begin{bmatrix} \cos(b) & \sin(b) & 0 & 0 \\ -\sin(b) & \cos(b) & 0 & 0 \\ 0 & 0 & \cos(b) & -\sin(b) \\ 0 & 0 & \sin(b) & \cos(b) \end{bmatrix} \right.$$

$$\begin{bmatrix} \cosh(c) & 0 & \sinh(c) & 0 \\ 0 & \cosh(c) & 0 & -\sinh(c) \\ \sinh(c) & 0 & \cosh(c) & 0 \\ 0 & -\sinh(c) & 0 & \cosh(c) \end{bmatrix}$$

$$\left.\begin{bmatrix} \cos(d) & 0 & 0 & \sin(d) \\ 0 & \cos(d) & \sin(d) & 0 \\ 0 & -\sin(d) & \cos(d) & 0 \\ -\sin(d) & 0 & 0 & \cos(d) \end{bmatrix}\right)$$

$$\exp\left(\begin{bmatrix} a & b & c & d \\ b & a & -d & -c \\ -c & d & a & -b \\ -d & c & -b & a \end{bmatrix}\right) = PROD \left(\begin{bmatrix} h & 0 & 0 & 0 \\ 0 & h & 0 & 0 \\ 0 & 0 & h & 0 \\ 0 & 0 & 0 & h \end{bmatrix} \begin{bmatrix} \cosh(b) & \sinh(b) & 0 & 0 \\ \sinh(b) & \cosh(b) & 0 & 0 \\ 0 & 0 & \cosh(b) & -\sinh(b) \\ 0 & 0 & -\sinh(b) & \cosh(b) \end{bmatrix} \right.$$

$$\begin{bmatrix} \cos(c) & 0 & \sin(c) & 0 \\ 0 & \cos(c) & 0 & -\sin(c) \\ -\sin(c) & 0 & \cos(c) & 0 \\ 0 & \sin(c) & 0 & \cos(c) \end{bmatrix}$$

$$\left.\begin{bmatrix} \cos(d) & 0 & 0 & \sin(d) \\ 0 & \cos(d) & -\sin(d) & 0 \\ 0 & \sin(d) & \cos(d) & 0 \\ -\sin(d) & 0 & 0 & \cos(d) \end{bmatrix}\right)$$

$$\exp\left(\begin{bmatrix} a & b & c & d \\ -b & a & d & -c \\ -c & d & a & -b \\ d & c & b & a \end{bmatrix}\right) = PROD\left(\begin{bmatrix} h & 0 & 0 & 0 \\ 0 & h & 0 & 0 \\ 0 & 0 & h & 0 \\ 0 & 0 & 0 & h \end{bmatrix}\begin{bmatrix} \cos(b) & \sin(b) & 0 & 0 \\ -\sin(b) & \cos(b) & 0 & 0 \\ 0 & 0 & \cos(b) & -\sin(b) \\ 0 & 0 & \sin(b) & \cos(b) \end{bmatrix}\right.$$

$$\begin{bmatrix} \cos(c) & 0 & \sin(c) & 0 \\ 0 & \cos(c) & 0 & -\sin(c) \\ -\sin(c) & 0 & \cos(c) & 0 \\ 0 & \sin(c) & 0 & \cos(c) \end{bmatrix}$$

$$\left.\begin{bmatrix} \cosh(d) & 0 & 0 & \sinh(d) \\ 0 & \cosh(d) & \sinh(d) & 0 \\ 0 & \sinh(d) & \cosh(d) & 0 \\ \sinh(d) & 0 & 0 & \cosh(d) \end{bmatrix}\right)$$

Glossary & Nomenclature

Algebra:
A linear space together with a multiplication operation that satisfies the requisite axioms.

Algebraic Closure:
An algebra is algebraically closed if every polynomial within it has roots that are also within the algebra. The euclidean complex numbers are algebraically closed.

Algebraic Field:
An algebra that satisfies the field axioms. A multiplicatively commutative division algebra

Algebraic Matrix Form:
A matrix form that together with matrix multiplication satisfies the requirements to be an algebra. See: Matrix Form

Algebraic Operations:
The operations of addition, multiplication, and multiplication by a scalar.

Angle Matrix:
A $n \times n$ matrix that a member of the SO_n group of matrices and that has simple-trigonometric functions or nu-functions (compound trig functions), or zero, for its elements. An angle matrix is the same as a rotation matrix.

Angle Product:
A way of multiplying two matrices together that is equivalent to both the inner product and the exterior product (s).

Cayley Table Standard Form:
The Cayley table of a group written with the top row in alphabetical order and the identity elements in the leading diagonal.

Compound Polar Form:
The form of a n-dimensional algebra written as a product of a length matrix and a single angle matrix. The angle matrix in the compound polar form is the product of $(n-1)$ angle matrices which have simple trigonometric functions for their elements. See: Simple Polar Form

Compound Rotation Matrix:
The single angle matrix in the compound polar form.

Compound-trig Function:
An element of a compound rotation matrix. They are projections from the unit sphere of the space directly on to an axis.

204

Diagonal Matrix:	A square matrix with all elements zero except the elements on the leading diagonal.
Dihedralions:	Division algebras whose under-lying group is a dihedral one.
Distance Function:	A function whose input is the co-ordinates of two points in space and whose output is the distance between those two points. It is invariant under transformation of axes and rotation.
Division Algebra:	An algebra in which one can always do division – there are no zero-divisors.
E-Type Algebraic Matrix Form:	An algebraic matrix form that gives rise to a simple polar form in which every angle matrix is a E-type angle matrix.
E-type angle matrix:	An angle matrix that has elements which are E-type simple-trig functions.
E-type simple-trig function:	A n dimensional simple-trig function. The series expansion has minus signs in it.
Exterior Product:	The exterior product of two vectors. It is related to the cross product of two vectors. It is denoted with a wedge.
Flat Algebra:	A natural algebra that has the space squeezing parameters $(j,k,l,....) = \pm 1$
General Polar Form:	The polar form of a general algebraic matrix form
H-Type Algebraic Matrix Form:	An algebraic matrix form that gives rise to a simple polar form in which every angle matrix is a H-type angle matrix. This algebraic matrix form has no negative elements (minus signs) in it.
H-type angle matrix:	An angle matrix that has elements which are H-type simple-trig functions.
H-type simple-trig function:	A n dimensional simple-trig function. The series expansion has no minus signs in it.
Hyper-geometric functions:	Messed about versions of the exponential function. A sub-set of the hyper-geometric functions is the simple-trigonometric functions.

Simple-trig
Function: The functions that occur in the simple angle matrices of the polar form. Combined together by matrix multiplication, they form the ν-functions.

Hyperbolic Complex
Numbers: One of the two flat 2-dimensional natural algebraic fields. The other is the euclidean complex numbers.

Inner Product: A projection function from one line in a space on to another line in that space. It is often called the dot-product or scalar-product. It is invariant under transformation of axes. See the inner product axioms in Appendix 1.

Leading Diagonal: The diagonal of a square matrix or square table that runs from the top left-hand corner to the bottom right-hand corner.

Length Matrix: A matrix that is isomorphic to a real number and that occurs in the polar form (simple or compound) of an algebra.

Linear Space: A set of mathematical objects together with an addition operation are a linear space if they satisfy the requisite axioms. A closed set of linear transformations.

Matrix Form: A matrix in which each element is specified either as independent of all other elements or related to them by a specified function. For example:

$$\begin{bmatrix} a_{11} = \mathbb{R} & a_{12} = \mathbb{R} \\ a_{21} = -5a_{12} & a_{22} = a_{11} + 2a_{21} \end{bmatrix}$$

Metric: A distance function that satisfies the metric axioms – Appendix 1.

n-dimensional
algebra: An algebra with n independent variables.

Natural Algebra: An algebra that has matrix multiplication as the multiplication operation.

Natural Algebraic
Field: An algebraic field that has matrix multiplication as the multiplication operation.

nu-function or
ν-function: A function in the compound polar form angle matrix. Also known as compound trigonometric functions. They are projection functions from the unit sphere on to the axes of the space.

Odd Parity Matrices: Matrices whose product is an element of a natural algebraic field.

Overblown Representation:	The matrix representation of an algebra by a larger matrix form than is necessary to contain the algebraic operations.
Permutation Matrix:	A matrix that is produced by swapping columns of the identity matrix. Each represents a permutation.
Pure Permutation Matrix:	A permutation matrix with an all zeros leading diagonal.

Polar Form of an
Algebra: The result of exponentiating the algebra.

$$\text{Eg: } \exp\left(a + \hat{i}b\right) = r\left(\cos\theta + \hat{i}\sin\theta\right)$$

Quaternions: The quaternion division algebra of William Hamilton. It contains the euclidean complex numbers as a sub-algebra.

Real Number Matrix: A matrix that has all elements equal to zero except the elements on the leading diagonal and in which the leading diagonal elements are equal. These matrices are algebraically isomorphic to the real numbers.

Rotation Matrix: *See: Angle Matrix*

Scalars: Real numbers or matrices that are algebraically isomorphic to the real numbers.

Shadow Algebra: A lesser dimensional representation of an algebra

Simple Polar Form: The form of a n-dimensional algebra written as a product of a length matrix and $(n-1)$ angle matrices. Each of the angle matrices in the simple polar form has simple trigonometric functions for its elements. See: Compound Polar Form

Simple Rotation
Matrix: A rotation matrix in the simple polar form of an algebra.

Simple-Trig
Function: A n-ways splitting of the exponential series.

Space Squeezing
Parameter: Parameters that appear in algebraic matrix forms which have the effect of squeezing the distance function 'out of shape'.

Sub-Algebra:	An algebra that is contained within a larger algebra. For example, the real number algebra is contained within the complex number algebras.
Study numbers:	Hyperbolic complex numbers
Trig Function:	Geometrically, a projection on to an axis.
Unit Sub-Matrix:	A matrix whose elements are either zero or 1 and which occurs as a sub-matrix of an algebraic matrix form – the b matrix with $b = 1$ for example. The multiplicative identity matrix is a unit sub-matrix of each algebraic matrix form.
Vector Matrix:	Another name for an algebraic matrix form. The term is used when considering the matrices as vectors.
Vector Space:	Another name for a linear space
Wedge Product:	The exterior product of two vectors.
Zero divisors:	A mathematical object, A or B, within an algebra is a zero divisor if $AB = 0$ when $A \neq 0$ & $B \neq 0$

208

Notation

$A_{[\text{Row, Column}]}$: Denotes a particular element of a matrix, A.

AH_n: The first H-type simple-trig function in n dimensions. This function has a power series expansion beginning with x^0 and containing no negative terms (minus signs).

AE_n: The first E-type simple-trig function in n dimensions. This function has a power series expansion beginning with x^0 and containing an equal number of negative terms and positive terms (minus signs & plus signs).

$BH_n...ZH_n$: The other H-type simple trigonometric functions in n dimensions. These functions have a power series expansion beginning with $x^{B...Z}$ and containing no negative terms (minus signs).

$BE_n...ZE_n$: The other E-type simple trigonometric functions in n dimensions. These functions have a power series expansion beginning with $x^{B...Z}$ and containing an equal number of negative terms and positive terms (minus signs & plus signs).

C_n: The cyclic group of order n.

$C_n \times C_m$: The group that is the direct product of the groups named

$C_{n.m}$: The name appelled to the general form of an algebra.

$C_{n.m} L^l H^p E^{(n-p-1)}{}_{(j=+1,k=+1,...)}$: The name appelled to a particular algebra. If the polar form is not known, the polar form part of the notation will be simply the length matrix.

$\mathbb{C}$: The complex numbers

det: The determinant of a matrix.

$\exp([A])$: The exponential of a matrix

$\begin{bmatrix} e & 0 \\ 0 & e \end{bmatrix}^{[A]}$: An alternative notation for the exponential of a 2 by 2 matrix

$\hat{i}$: The square root of minus unity - $\sqrt{-1}$. In general, such objects will carry a hat symbol above them.

$L^l H^n E^m$ The polar form of an algebra

$\log m_{[b]}([A])$: The logarithm of a matrix to base $[B]$ where B is a matrix that is algebraically isomorphic to the real number b and of the same size as the matrix A.

$v_{[\ ]}A(b,...z)$: The nu-functions in the compound polar form angle matrix of the algebra that appears in the subscript. They correspond to the cosine or the cosh function in 2-dimensions. They are sums of products of simple trigonometric functions formed by the matrix multiplication of the $(m+n)$ angle matrices in the simple polar form.

$v_{[\ ]}B...Z(b,...z)$: The nu-functions in the compound polar form angle matrix of the algebra that appears in the subscript. They are sums of products of simple trigonometric functions formed by the matrix multiplication of the $(m+n)$ angle matrices in the simple polar form.

$\mathbb{OP}_c$ The odd parity matrices associated with the euclidean complex numbers
$\mathbb{OP}_s$ The odd parity matrices associated with the hyperbolic complex numbers

$PROD(\ )$:   The product of the objects within the bracket.

$\hat{p},\hat{q}$:   The cube roots of plus or minus unity in the non-matrix notation of the 3-dimension algebras - $a+b\hat{p}+c\hat{q}$. This is analogous to the non-matrix representation of the complex numbers.

$\mathbb{Q}$:   The quaternions of William Hamilton
$\mathbb{R}$:   The real numbers
$\mathbb{S}$:   The hyperbolic complex numbers (Study numbers)).

$\{\triangleright,\triangleleft\}$ Used to indicate equivalence under reduction from the overblown representation or from an algebraically isomorphic representation. These symbols are also used to indicate the folding and unfolding operations.

$[A]\odot[B]$ The angle product of the matrices $\{A,B\}$. The angle product contains the vector inner product and the exterior vector product(s).

$\begin{bmatrix} a \\ b \end{bmatrix} \wedge \begin{bmatrix} c \\ d \end{bmatrix}$ The exterior product of two vectors. Also known as the wedge product.

Index